P9-BXY-469

Van Buren
District Library
Buy
a
Book
Program

Presented to the
ANTWERP SUNSHINE
LIBRARY

Barbara Farris

DISCARDED

GREAT
MIGRATIONS

GREAT MIGRATIONS

OFFICIAL COMPANION TO THE NATIONAL GEOGRAPHIC CHANNEL GLOBAL TELEVISION EVENT

K. M. KOSTYAL

AFTERWORD BY SERIES PRODUCER

DAVID HAMLIN

NATIONAL GEOGRAPHIC

WASHINGTON, D.C.

CONTENTS

MIGRATION MAP — **6**

INTRODUCTION — **16**
Restless Planet

CHAPTER 1 — **24**
Born to Move

CHAPTER 2 — **78**
Need to Breed

CHAPTER 3 — **148**
Race to Survive

CHAPTER 4 — **214**
Feast or Famine

EPILOGUE — **282**
The Future of Migration

PRODUCER'S NOTEBOOK BY DAVID HAMLIN — **292**

Call to Action **294**
About the Film **296**
Bibliography **298**
About the Authors **300**
Gallery Captions **301**
Photo Credits **302**

MIGRATION MAP

The following animals are featured in the stories that begin on the page numbers listed below.

108   Army Ants

260   Bald Eagles

90   Black-Browed Albatrosses

260   Canada Geese

90   Caracaras

180   Gibbons

250   Golden Jellyfish

224   Great White Sharks

236   Mali Elephants

260   Mallard Ducks

260   Mayflies

58   Monarch Butterflies

224   Northern Elephant Seals

180   Orangutans

204   Pacific Walruses

260   Peregrine Falcons

OCEAN

PACIFIC
WALRUSES
Chukchi Sea

PACIFIC
WALRUSES
Bering Sea

EUROPE

ASIA

AFRICA

NORTH

PACIFIC

OCEAN

GOLDEN JELLYFISH
Palau

WHITE-EARED KOB
Sudan

SERENGETI ZEBRAS
Kenya and Tanzania

WILDEBEESTS
Kenya and Tanzania

ORANGUTANS
Borneo, Malaysia and Indonesia

SPERM WHALES
Pacific Ocean

RED CRABS
*Christmas Island,
Australia*

INDIAN

OCEAN

ZEBRAS
Botswana

RED FLYING
FOXES
Australia

AUSTRALIA

SOUTH PACIFIC

OCEAN

SPERM WHALES
Indian Ocean

0   1000   2000   3000
KILOMETERS

0   1000   2000   3000
STATUTE MILES

Scale at the Equator

ANTARCTICA

Phytoplankton 172

Pink Salmon 128

Proboscis Monkeys 180

Pronghorn Antelope 192

Red Crabs 48

Red Flying Foxes 138

Rockhopper Penguins 90

Snow Geese 260

Sockeye Salmon 128

Southern Elephant Seals 90

Sperm Whales 70

Tundra Swans 260

Whale Sharks 172

White-Eared Kob 120

Wildebeests 34

Zebras 160

NOTE ABOUT THE MAP

This map generally depicts the locations of animals featured in this
book. It does not necessarily represent the species' entire habitats
or migratory ranges.

It may be the greatest spectacle nature orchestrates—a story of hunger and thirst, birth and death, violence and triumph. Migration is the ultimate drama—the elemental story of instinct and survival. For without the great migrations—the waves of wildebeests crossing the Serengeti, the wing beats of a billion monarchs lifting from the mountains of Mexico, the armada of Pacific walruses floating with the ice through the Bering Strait—whole species would head toward extinction.

RESTLESS PLANET

But what internal and external forces drive animals to make these risky yet deliberate journeys, crossing hundreds, sometimes thousands of miles? What tells them when to move, and what guides them to their destinations? What propels them to face predators and natural forces that are sure to cost the lives of many of them, particularly their young?

These are the questions that scientists have worked for decades to answer. They have some of the answers for some of the animals, but the great migrations remain an inspiring mystery of navigation and endurance—a mystery that is mulled and celebrated in the pages of this book and in the National Geographic Channel series that it echoes.

The world of animals has long been a cherished topic at National Geographic, and its magazines, books, and films have captured groundbreaking images of the

A MALE WANDERING ALBATROSS displays its 11-foot wingspan to a female on South Georgia Island. The courtship ritual renews their pair bond after months of roaming the Southern Ocean.

intimacy and subtlety of animal behavior. One of the Geographic's first television programs, *Miss Goodall and the Wild Chimpanzees*, became a classic when it aired on CBS in December 1965. As cinematography grew more sophisticated, so did the programming. Following the Goodall special, the Geographic produced *Grizzly, Winged World, The Mystery of Animal Behavior*, and more recently, the acclaimed *March of the Penguins* and *Arctic Tale*.

Such cinematographic tours de force would no doubt have amazed the group of prominent scientists and academicians who met in Washington, D.C., in 1888, hoping to establish an American society patterned after Britain's Royal Geographical Society. Their brainchild, the National Geographic Society, was dedicated to the "increase and diffusion of geographical knowledge." To that end, they began publishing an unillustrated magazine aimed mostly at scholars and explorers. Today, the Society they founded has gone far beyond their wildest dreams—taking the diffusion of geographic knowledge in directions and into media they could never have imagined.

In ways they, and we, can only fathom, migration is a key to the intricacy of life on Earth, creating webs and networks whose complexity and fragility we are just beginning to grasp. New technologies promise to help both scientists and the public better understand how animals move and what they need to sustain their migrations. Satellite tracking is one of the most recent such innovations, allowing scientists to follow animal movements in ways that were impossible only a few decades before.

MONARCH BUTTERFLIES FLOAT through a forest like confetti in their winter home in Mexico after migrating as far as 2,500 miles. Immense numbers—100 million to 350 million, by recent estimates—gather in mountain forests, where the cool climate favors their metabolism. Some 138,000 forested acres are set aside as a protected biosphere reserve, but logging threatens this critical habitat outside the reserve.

New transmitters small enough to attach to the four-ounce arctic tern have revealed that the intrepid flyer covers 44,000 miles in its zigzagging migration between Greenland and Antarctica—the longest distance covered by any migrating animal. Satellite tracking has also followed white sharks on their annual 6,000-mile pilgrimage across the North Pacific and caribou making a 3,700-mile annual trek between the forests and the tundra of northern Quebec. These findings have overturned long-held beliefs about the animals' habits and movements.

The new tracking technologies often go where humans can't follow, but for many biologists, their most significant work is still done the old-fashioned way. They spend months, years, even a lifetime in the desert, at sea, on the tundra, finding animals by sight and sound, then observing, recording, and trying to fathom nuances known only to the animals themselves.

Thanks to the diligence and dedication of the scientific community, the public at large now has a much better understanding of the great migrations and a deeper appreciation of how critical they are to the biodiversity of the planet. Fewer salmon returning annually to Alaska waters, for instance, means that many other creatures will suffer, from apex predators like the salmon shark to the forests themselves to the zooplankton in rivers—all of whom rely on the spawning salmon in one way or another.

Even the movements of animals that make daily migrations may have the most profound impacts. Scientists studying a rare marine lake in Palau believe that the daily pilgrimage of golden jellyfish up, down, and across the lake may mix and churn

A WALRUS CALF huddles with its mother as great herds assemble to migrate through the Bering Strait, drifting northward on ice floes to the Chukchi Sea in the spring and paddling back southward to the ice-free Bering Sea for the winter.

the waters in ways that breathe life into them. And other marine biologists speculate that the ocean's deep scattering layer—a collection of migrating zooplankton—may be sequestering carbon, with critical implications for the global climate.

But migration itself, as a grand global phenomenon, is threatened by a host of new and shifting conditions. Not only is climate change creating its own challenges, but the habitats needed for migration corridors are being fragmented and degraded by human activities—by more settlements, roads, farm fields, livestock, suburbs, and exurbs. From the grasslands of Mongolia, where the threatened saiga antelope finds its seasonal paths narrowed, to the plains of Wyoming, where pronghorn have to cross highways to pursue their age-old route to the mountains, long-distance migrants are being stressed.

Without question, migration impacts more than the migrants. The vast roaming herds of bison once aerated and fertilized the Great Plains of America, and greater numbers of warblers than today once helped to keep insect populations in check. In ways beyond our understanding, migrations may be critical mechanisms for holding the planet in balance.

In the pages that follow, the seemingly choreographed miracles of movement that take place across the globe every hour, every day, and in every season become more than scientific phenomena. They become the common drama of survival that unites all earthly inhabitants.

A WILDEBEEST HERD descends a rock outcropping during migration across Tanzania's Serengeti Plains. The herd moves almost continuously through the year, tracking seasonal rains and grasses in a slow, clockwise circle of more than 1,800 miles. The migration is interrupted for only two to three weeks in late fall, when the females give birth to a new generation of calves.

BORN

Whether on the air, in the sea, or across the land, the great migrations are always odysseys of survival. The risks involved can be overwhelming, but running the risks seems bred in the bone. For many animals, the migrations are seasonal pilgrimages undertaken for the most elemental of reasons—to reproduce their own kind. The **CHRISTMAS ISLAND RED CRABS** make that kind of journey to the sea annually—even though the whole arduous process is usually futile. **MONARCH BUTTERFLIES** take to the wing for the

TO MOVE

same reason, beginning a long, multigenerational migration that is the only way to ensure their survival as a species. • For other animals, motion is life, a perpetual journey that never ends. Male **SPERM WHALES** ceaselessly roam the deep oceans, loners whose solitude s broken only by the periodic need to connect with others of their own kind. The **WILDEBEESTS,** too, are ceaseless roamers, running for their lives across the Serengeti, chasing the rains that take them n an endless loop of survival, migrants whose migration never ends.

They surge across the Serengeti-Mara like the force of nature that they are. More than a million wildebeests, careering around the endless open plains and acacia savannas of Kenya and Tanzania. Their gangly shapes, made for running, are at odds with their bearded mandarin-like faces, which seem to hold an acceptance of all things. Their lives are literally a continual race for survival, a long migration that never ends. The herd is their life, a creature in its own right, regulating each individual animal's movements, mating, birth, even death.

NOMADS IN THE LAND

A convergence of sun, wind, rain, and geology fuels the wildebeests' annual loop through the land they have occupied for 1.5 million years. Each year, they travel some 1,800 miles, churning up the soils of equatorial East Africa in a race toward rain and the green it engenders. Their journey has no beginning and no end, but on calendar time, from December through the first few months of every new year these great herds of herbivores mass at the south end of Serengeti National Park and the adjacent Ngorongoro Conservation Area. Sometimes as many as 2,500 animals crowd onto a square mile, all grazing on the green bounty brought on by the "short rains." During a brief two or three weeks the females in this enormous aggregation of animals give birth to the new generation of calves—as many as a half million of them.

The sheer number of young keeps the vast herds vast, because only one out of every six of the newborns will survive its first year. Within minutes of birth, the calves are on their feet, finding a tentative but crucial foothold for their coming life on the move. Motion and the mass of the herd are what will keep them alive, particularly in their first vulnerable year. However the wildebeest mothers manage to communicate it, the message is clear: Keep moving and keep your head down. Meld into the herd, don't call attention to yourself, because the plains predators—lions, cheetahs, hyenas, and wild dogs—are always waiting.

Wildebeests are remarkably fast animals, sprinting at speeds of up to 50 miles an hour, but their young especially are no match for the cheetah, whose bursts of speed propel it toward its prey at more than 75 miles an hour, faster than any animal on land. And the cheetah is only one of the threats the herd faces.

In the shade of the kopjes, the rock outcroppings that punctuate the short-grass prairie, lions wait out the daylight with their prides, preferring to do their hunting at night. Under cover of darkness, the lions make their move. Crouching as they inch closer to the grazing herd, they explode out of the night, racing toward the most vulnerable targets—an ailing adult wildebeest, or a calf or a yearling that has wandered too far from the herd. The wildebeests take off in a spray of hooves as the big cats sprint after them, grabbing at haunches with

WILDEBEEST—"WILD BEAST," as Dutch settlers named them; gnu, as they are called in a derivation of an African-language name—race across the dusty plains of Serengeti National Park in Tanzania at sunset.

A CHEETAH WALKS past a grazing wildebeest and zebra herd
in Maasai Mara National Reserve, Kenya.

The predator's target is often a newborn wildebeest taking its first unsteady steps,
such as this one in Serengeti National Park, Tanzania, whose mother guards it closely.

their teeth and claws. Once they've made the kill, the lions eat their fill. Hyenas and vultures will scavenge what they leave behind.

The wildebeests do have one strategic advantage. They—including their young—are built to stay on the move, while the dependent young of lions are more or less sedentary. The lions have to wait for the wildebeests to wander into their territory before they attack. And the short-grass plains offer the hunted another strategic advantage—good

visibility. It's easy to spot a stalking predator here than in the higher grasses. The short grass, if not the greenest, has another great benefit for the wildebeests—it's rich in minerals critical to their survival.

Eons ago, in the Pleistocene, volcanoes laid down the ash that formed the *siringet*—in the language of the local Maasai "the place where the land runs on forever." In the southern Serengeti, rains leached the calcium out of the ash, creating a dense hardpan below a thin layer of topsoil. Only short grasses with shallow roots can grow here, but the grass is rich in phosphorus, thanks to the ashy soil. As the mother wildebeests feed, they take in

the phosphorus critical to their own milk production and to the development of their young.

As they graze their way up the Serengeti, the wildebeests aren't alone. Some 200,000 plains zebras and 350,000 Thomson's gazelles are part of a vast natural network, a complicated convergence of flora and fauna, weather and geology. The zebras stay in the vanguard of this movable feast. Voracious grazers, they have teeth that can tear at the dead, coarse tops of the plains grasses, exposing and enabling new leaves and stems to grow.

Following behind, the broad-muzzle wildebeests graze on the new growth they need to survive. Like a massive eating machine, they plow through the short-grass plains, leveling them and at the same time fertilizing them with their dung and watering and refortifying them with their saliva and urine. Their hooves break up the soil, encouraging

regeneration. By the time the Thomson's gazelles bring up the vanguard in the wildebeests' wake, the Serengeti is carpeted in green again.

Making a slow clockwise arc on the annual migration, the wildebeests stay on the move, following each other, often single file, guided by a pheromone trail left in the grasses by the hooves of those ahead of them in the herd. Constantly on alert for the rain that soaks and brightens the Serengeti from March to May, they move toward lightning or thunderheads on the horizon or toward the smell of moisture and growing humidity. They know the plains will green up after a rain, and waterholes will be replenished. Predators may be waiting for them at the oases, but they have to take that chance.

By late May or early June, when the rains come to an end, the wildebeests move to the western corridor of the park, and the annual rut begins. All of the mature females go into estrus during this time, ensuring that they'll give birth after the short rains have greened the plains again. As the rut begins, males square off to establish territory, and they begin displaying, hoping to attract females to their territory. The fights between bulls are little more than ritualized posturing, and the females wander freely among the territories, mating with a number of males.

Like calving, the rut is compressed within a few brief weeks, and then it's over. As the sun parches the African earth, the wildebeests move quickly northwest, anxious to leave Tanzania's water- and

The crocs take only a small percentage of the herd compared to the river. Yet despite its perils, the wildebeests are utterly reliant on the Mara, the only year-round water source in the region. In recent years the reliable flow of the Mara has diminished as its watershed, the Mau Forest in the Kenya highlands, has been cut down by charcoalers and by farmers clearing the land for wheat fields. Without the moist, spongy forest floor to feed it, the Mara could dry up, and without the river, predicts scientist Robin Reid, "the wildebeest population would collapse."

That's not the only threat to the great migration. Poaching for the bushmeat trade is estimated to claim as many as 200,000 animals a year, and even more critical, the very lands required for the wildebeest migration are at risk. While part of their long migratory loop is protected in national park and reserve lands, the wildebeests frequently range beyond these areas. As a rising tide of humans pushes up to the edge of those preserves, more and more farm fields and fences stand in the path of wildebeests' ceaseless wanderings.

grass-poor Serengeti behind. They're heading at a trot for the acacia savannas of Kenya's Maasai Mara National Reserve, where food and water will ensure their survival through the dry season.

But to get there they have to undergo perhaps their greatest challenge—the Mara River. For reasons unknown, the wildebeests head for only a handful of fording points on the river. The river itself, fast and relentless, sweeps thousands of them away, leaving their bodies bloated and massed along the riverbanks. Waiting for them at the crossing points are predators they can't outrun.

Twenty feet long and weighing in at some 1,500 pounds, a Nile crocodile can bring down a 500-pound adult wildebeest. Clenching its massive jaws around the equally massive head of its prey, the crocodile immobilizes it and drags it to its death in a spectacle of sheer animal force. It can devour half its own body weight in one feeding, and the crocodiles in the Mara plan their own annual cycle around the feeding frenzy promised by the wildebeest migration.

In the late 1950s the director of the Frankfurt Zoo, Bernhard Grzimek, and his son Michael began making flights above the Serengeti in order to get a bird's-eye view of the great migration and some sense, at last, of the wildebeests' true numbers. Their groundbreaking aerial surveys cost Michael his life. In 1959, during the filming of *Serengeti Shall Not Die,* Michael's small plane collided with a vulture and crashed. Still, the film became a classic, propelling the public toward a greater sympathy and understanding for the fragility of Africa's natural world.

WILDEBEESTS FACE PREDATORS such as the cheetah (opposite, top; above right).
The cheetah's method is to grasp a leg and toss the animal to the ground (opposite, bottom).
Constantly watching, vultures never miss an opportunity to feed on what's left (above left).
Following pages: A wildebeest herd stampedes across the dusty plains of Maasai Mara National Reserve in Kenya.

In the book by the same name Grzimek wrote: "Only Nature is eternal, unless we senselessly destroy it. . . . When, fifty years from now, a lion walks into the red dawn and roars resoundingly, it will mean something to people and quicken their hearts. . . . They will stand in quiet awe as, for the first time in their lives, they watch twenty thousand zebras wander across the endless plains."

Grzimek well knew the destruction humans could wreak, even unintentionally. When he arrived on the Serengeti, the wildebeests were still in the thrall of a plague that had decimated their numbers. Sometime in the 1880s, rinderpest had come into eastern Africa via cattle imported from Europe. In a few short years, it claimed 95 percent of the wild buffalo and wildebeest population, and the same percentage of the domesticated cattle that East Africa's Maasai and other nomadic herders and pastoralists relied on to survive. Lions that typically feed on large hoofed mammals became man-eaters. And without the grazers and browsers that moved ceaselessly across the

Serengeti, woodlands and thickets began to replace grasslands in the northern plains, where the soils are deeper and the rainfall more plentiful. Finally, after 70 years, immunization conquered rinderpest in cattle, and that in turn halted it in wildebeests. From a low of 220,000 in 1961, the wildebeests numbers swelled to 1.4 million animals in 1975. Today, the great herd that sweeps through Kenya and Tanzania numbers somewhere between a million and 1.2 million animals.

This annual pilgrimage, one of the largest land-animal migrations left on the planet, is predicated on open space. Without the room to roam with the rains, the wildebeests can't thrive. Conservation groups and governments are working hard to preserve their migratory corridors, but in a land of scant resources and ever rising human demand, it's an uphill battle. Meanwhile, the wildebeests, oblivious to watersheds, park boundaries, and the geopolitics that could save or doom them, are on the move, racing with the rains across the land that runs on forever.

PLAINS ZEBRAS typically accompany the wildebeest, and danger lurks if the big equines become separated from the herd. They are targets of attack by lions and must be both vigilant and fast to escape (above). Individual zebras find safety amid the tightly packed wildebeest migrants (opposite). Following pages: Wildebeests cool themselves in a Tanzanian river.

DANGER LURKS when wildebeests cross the Mara River in Kenya, where Nile crocodiles lie in wait.

When a crocodile leaps up from the water and strikes,
the wildebeests instantly dash through the river in panic.

Christmas Island lies about 220 miles south of Java, an almost indiscernible dot in the great sweep of the Indian Ocean. But every November, when monsoons cloud the sky and humidity seeps into the land, the skin of the island begins to boil red and come alive, seething with millions of creatures determined to make a migration that will probably end in defeat.

MARCH TO THE SEA

Christmas Island's fabled red crabs, *Gecarcoidea natalis,* live solitary lives almost all year, confining themselves to the rain forests that cover the island's central plateau. They scurry across the forest floor like vacuum cleaners, devouring its leaf fall, flowers, and seedlings. Their leavings act as fertilizer, recycling nutrients, and their burrows help till and aerate the soil. As the dry season sucks the air clear of the humidity that these land crabs desperately need, they hole up in solitary burrows, plugging them with wads of leaves to trap the moisture. And there they wait for two or three months.

With the first hint of rain in late October or November, they resurface and begin their annual pilgrimage to the sea. The rains surely help trigger their urge to migrate, but hormones as well seem to play a role. One particular hormone released during the wet season encourages the production of invigorating glucose. For the migration ordeal ahead, the red crabs will need all the energy they can get.

Negotiating the limestone terraces that ring the central plateau and give the island its wedding-cake shape, the crabs head for the shore terraces along the northwest coast, several miles and 700 precipitous feet below. It's unclear exactly what guides the red crabs, but scientists believe it's probably visual cues and a route learned in the past, as well as polarized light and possibly the crabs' own magnetoreceptors.

The big males, several inches across, go first. The oldest ones have done this for a dozen years, and no matter what niche they occupy on the central plateau, they know the most direct route to the sea, to the very beaches where they themselves were born. They pace their march to the cool of the day, moving in the morning and late afternoon. Heat is the enemy, quickly dehydrating them. They instinctively know to avoid heat, but automobiles and trucks are harder to combat, even though the government has posted "crab crossings" and constructed tunnels for them beneath roads along their migratory route. Golfers on the local course (built, for some reason, right in the middle of the crabs' annual path) also know that the roiling red crustaceans have the right of way.

Even more fatal to the crabs than vehicles is a relative newcomer to the island—the yellow crazy ant, a tramp species native to Africa that has invaded some Pacific islands. Ferocious and systematic killers, the ants spray formic acid into a crab's eyes and mouthparts, and then attack it en masse. And they don't just feed on crabs. Highly social animals that cooperate together and form supercolonies, the ants even attack larger animals. The red crabs are no match for them. In just three decades they have decimated the crab population, leaving it half its former size.

BEFORE MIGRATING to the seashore to mate and spawn, red crabs live alone in moist burrows during the arid season until rains trigger a release of energy-building hormones.

When the crabs emerge and migrate to the seashore to mate,
their perils include crossing railroad tracks and climbing down steep seaside cliffs.

More than affecting just the red crabs, the ants have begun to impact the whole island ecosystem. They also prey on scale insects that live in the rain forest, and the way these insects themselves feed stimulates the growth of sooty molds that in turn lead to the death of trees. Red crabs, too, have had a somewhat negative affect on the forest, sucking up seedlings that would regenerate to form a forest understory. Thanks to the predatory ants, the forest may actually benefit from fewer crabs and fewer scale insects. But ecosystems are complicated and delicately balanced organisms. When their processes shift and change, the outcome is hard to predict.

In the meantime, the red crabs continue their long march to the sea. Most average something less than half a mile a day, but how quickly they move is dictated by the rains. If the monsoons arrive in a timely fashion, then the migration is less frenzied, and the crabs can move at a more leisurely pace, feeding along the way. But when the rains are delayed, as they have been in recent years, the pilgrimage becomes a race to the sea. Without the heavy rains, these land crabs, so reliant on ambient moisture, die of dehydration and exhaustion before they reach the sea.

In an optimal season it takes the first males about a week to reach the shore terraces on the northwest coast, where the ocean is calmer. The first thing the males do is go down to the sea to soak up life-giving moisture and salts from the surf, tide pools, wet sand, and rocks. Once replenished, they move back up to the shore terraces, where they frantically dig mating burrows sometimes so close that there are two burrows on every square meter of terrace. The fighting for territory can be fierce, at times even mortal. In a day or two, the females arrive. After their own regenerative dip in the sea moisture and salts, they are lured by males into their burrows to mate.

Once the coupling is accomplished, the males leave the burrows behind and begin the reverse migration, heading upland to the island's forested center. The female red crabs remain in the burrow, and three days after mating each female produces eggs, some 100,000 of them. For 12 more days she holes up in the protective moisture of the burrow, as the eggs in her brood pouch mature.

Then, within a day or two of the last quarter of the moon, the "berried"—egg-heavy—females begin to emerge. It's during this lunar phase that the sea level fluctuates least between high and low tide, and all the steps in the red crab migration are timed to this. In the limited shade above the waterline, the female crabs literally pile on top of one another, forming flaming red heaps that give off an eerie squeaking, like the cry of a young bird. In the darkness of night, at the turn of the high tide, the mothers begin a final cautious crawl

DURING THEIR WEEKLONG MIGRATION, red crabs must climb down precipitous cliffs (top left), where many fall and die before reaching their destination on the shore (above). There, they hurry toward the badly needed moisture of the surf and tide pools before mating begins (opposite).

FEMALE RED CRABS with full brood pouches descend a cliff in the moonlight to spawn, releasing their eggs into the ocean.

Only a very small portion of newly hatched crabs make it back to land.
Most are eaten by ocean predators or swept away in the sea by strong currents.

to the sea that crescendos into a reckless rush into the surf. Dancing what looks for all the world like a jig and flexing their abdomens to break open the eggs, they release skeins of tiny larvae into the ocean. Night after night, for five or six nights, the egg-releasing spectacle plays on. Some females try to release their eggs from cliff faces 20 feet above the sea. Some succeed in the tricky maneuver, others fall to their deaths trying.

But that's only one tragedy in this often tragic migration. Whale sharks, manta rays, and other fish gobble the red larvae by the millions as tide and surf carry them farther out to sea. The larvae have to survive about 25 days in the inhospitable ocean before they develop into tiny prawnlike, big-eyed megalops. The ones that make it to this stage and make it back to the island spend another day or two near shore, where they take on the look of a tiny crab.

Now, no more than a fifth of an inch across, the survivors finally emerge from the sea. If there are survivors. Most years, there are none—and the entire dangerous ordeal of the red crab migration has been an exercise in futility. But every one or two years in ten, millions of the tiny crabs make it onto land, crawling from the ocean like a conquering army, invading human homes and anything else in their path as they make their way instinctively up to Christmas Island's central plateau. Their sheer numbers are enough to guarantee the survival of the species—that is, if herons and thrushes, climate change, yellow crazy ants, and the antics of humans don't one day end the red crabs' remarkable march to the sea.

RED CRAB LARVAE that survive marine predators are a tiny proportion of eggs laid. The larvae soon become crablike (above) and begin their first trek to the forest (right).

The monarch migration ranks as one of the great annual epics, a generational odyssey played out on prairies and plains and the forested slopes of Mexico's volcanoes. The potential dangers are classic: hunger and thirst, freezing temperatures, even cannibalism. The heroics are impressive—a creature weighing less than a hundredth of an ounce undertaking a migratory flight of more than 2,000 miles. And yet those spotted orange monarchs that waft across our backyards and farm fields seem to belie the high drama and perfect species coordination that it takes them to survive.

MARIPOSA MIRACLE

For all monarchs, the drama begins under milkweed. To them the plant is their lifeline, serving as an incubator for their eggs and the virtual mother's milk that will nourish the newly hatched larvae. Without it, the monarch young, and the species, would not survive.

In the early Texas spring, as the new milkweed crop begins to sprout, pregnant female monarchs suddenly appear, flying up from Mexico. The females need as much milkweed as possible for successful egg laying. Each mother has roughly 200 eggs to deposit, and optimally, the eggs should be laid separately, among a number of milkweed plants, as a protection against sibling cannibalism. For four days, the larva inside the egg matures. Then the caterpillar chews its way out of its eggshell and begins to inch along the milkweed. It will happily devour any unhatched monarch egg in its path, but the milkweed is its true meal and protection. Since the plant's milky essence, its latex, is full of glycosides that are poisonous to other animals, the milkweed-sated monarchs are unpalatable to most potential predators.

The caterpillar spends its first two weeks feeding ravenously on milkweed, adding enormous bulk to its body. Finally, it sheds its larval skin and transforms itself into a green chrysalis hanging from a silk thread under tree limbs, leaves, even the eaves of buildings. Inside this jade nugget the monarch is literally dissolving into itself, its cells morphing until the transformative miracle is complete. In the final day, the form of the butterfly takes shape within the now almost transparent pupa. Then in a brief minute the fully formed butterfly emerges from its casing.

Following the spring and the milkweed sprout up the plains, the newly hatched monarchs begin their own life cycle. They seem to track the weather, looking for the mild temperatures that will encourage their reproductive chances and the development of their larvae. As the season progresses, the older generations of monarchs end their journey, but new generations continue the annual pilgrimage northeastward, some making it all the way to southern Canada.

A MONARCH'S TOXIC LIFE begins when a caterpillar feeds on a milkweed containing a poisonous chemical with a disagreeable taste that predators soon learn to avoid.

During the second stage the pupa in a chrysalis develops gradually toward adulthood,
finally emerging and ready to fly.

Dosed with milkweed toxicity and outfitted in bright color, always a warning sign to predators, the monarch is generally well fortified against predation. But not all varieties of milkweed it feeds on are toxic, and predators sometimes strike. The carnivorous praying mantis, deceptively still and unaggressive in its hunting pose, will grab an unsuspecting monarch with lightning speed and make a quick meal of it.

As fall begins to cool the air, the fourth generation of monarchs goes into a reproductive hiatus, storing its energy for the final challenge in the migratory epic—the long sweep south to the ancestral home in Mexico. With the sun as a compass and the Earth's magnetic field as a navigational tool, the butterflies unerringly return to a small sweep of forest in Michoacán, a daunting 2,000 miles away.

Nectar sustains this last long leg of the migration. Gathered from aster and goldenrod, ironweed, clover, and alfalfa, the nectar is converted into lipids, or fats, that can be stored to fuel the flight south and to see the monarchs through the winter. But there is a carrying charge, an energy cost, for that fuel load. Those costs can be partially offset by the wind. When the right winds are blowing, monarchs

catch them and soar almost effortlessly forward; without the winds, every wing beat has a price.

How quickly the butterflies cover the return miles depends on wind and weather, on the availability of nectaring plants, and no doubt on other variables as yet unknown. While the speed during the return migration can vary a lot, the median seems to be about 28 miles a day. Impressive for a creature so delicately designed as to seem virtually weightless.

By winter the monarchs have made their way back to the forested slopes of the Transvolcanic Range, about 60 miles

west of Mexico City. Hundreds of millions of them settle into a dozen colonies in the old-growth *oyamel* fir and pine forests on remote volcanic slopes where they'll wait out the winter. Hundreds of millions of brightly colored insects would seem easy enough to spot, and yet the monarchs' winter home remained a mystery until the mid-1970s. Before that, most scientists assumed the butterflies wintered in tropical or subtropical environments, but exactly where they went no one knew.

Then in the mid-20th century Canadian entomologist Fred Urquhart began a long and systematic attempt to track the monarchs' movements. He enlisted volunteers to tag the wings of individual butterflies with a tiny piece of white adhesive that read, "Return to Museum, Toronto, Canada," and he used the returned specimens to plot the monarchs' movements on a map.

Urquhart's methodical diligence paid off in stupendous ways. He first discovered that there are two populations of monarchs in North America. A western population lives west of the Rockies and over-winters along the mid-California coast—most famously in Pacific Grove, on Monterey Bay. A much larger eastern population graces

A PRAYING MANTIS (above) may be immune to a monarch's toxin and hungrily devours one (top right). The toxin may not harm spiders either, and they readily capture adults (opposite).

the springs, summers, and early falls of the American plains and the East. But where, *where* did this great eastern mass of insects go in the winters?

One of Urquhart's tagged monarchs was recovered in Mexico, but one butterfly could provide no more than a tantalizing clue. Still Urquhart and his wife, Nora, pursued it, traveling to Mexico to interview and enlist the help of everyone from school kids to amateur enthusiasts. Had anyone seen great numbers of *mariposas?* No one had. Then in 1974, Ken Brugger, an American engineer living in Mexico City, joined the chase, cranking his motorcycle into the mountains beyond the city. He found that the small town of Angangueo seemed a locus of monarch activity, and his finding enticed the Urquharts to the high mountains.

Occasional snow swirled around them, and the 10,000-foot altitude left them short of breath as their Jeep climbed higher on a raw winter day in 1976. Could this inhospitable spot have anything to do with monarchs? Then Urquhart saw them: "Butterflies—millions upon millions of monarch butterflies! They clung in tightly packed masses to every branch and trunk of the tall, gray-green *oyamel* trees. They swirled through the air like autumn leaves. . . ." So dense were they that they "blocked out the light," allowing only a bit of sun to illuminate the forest floor, "which looked like a gigantic Persian carpet because it, too, was covered with orange monarch butterflies." Urquhart had solved the great mystery. This was where the monarchs overwintered. Now the mystery was why.

MONARCHS CLUMP by the millions on oyamel trees
in Mexican forests (above). Before migrating northward in the spring,
the butterflies drop from the trees and begin a giant mating spree (right).

The answer has a Goldilocks simplicity. Not too hot or too cold, the temperatures in the Mexican winter forests are just right for the monarchs during their months of dormancy. The cold but not fatally freezing temperatures slow metabolism, allowing the cold-blooded monarchs to burn their lipid reserves more gradually. The greatest threat to them in the winter months comes from birds that prey on them, particularly when they rouse themselves to drink from nearby streams to avoid dehydration.

In mid-February, monarchs become more and more active, the males beginning the mating ritual. By March the pairing of millions of monarchs has reached a frenzied pitch. Then in early spring, rising up from the forests, clouds of monarchs follow the prevailing winds northeast, reaching Texas in late March and early April, in time for the milkweed sprout and the maturing of the first generation of eggs. And under milkweed, the multigenerational, 4,000-mile saga begins again.

But like so many eons-old migratory odysseys, the monarchs' is in jeopardy. Their winter home in the Mexican forests is threatened by illegal logging and an encroaching human presence. And the milkweed diet their young rely on is also at risk. Corn pollen from transgenic hybrids blowing onto the milkweed can kill feeding monarch larvae. And common milkweed, *Asclepias syriaca,* which the monarchs prefer, has declined dramatically in recent decades, a victim of the herbicides sprayed across farm fields, particularly in the Midwest, and human development, which has eliminated open fields where milkweed thrives. Today, milkweed and monarch are poised in a delicate balance.

One of the world's leading expert on monarchs, Lincoln Brower, fears that the migration of the eastern population may be "an endangered biological phenomenon." Will the spectacle of millions of butterflies filling the sky go the way of the passenger pigeon migration and become a thing of legend? Or will the great monarch migration remain one of the most dazzling feats of the natural world?

MIGRATING MONARCHS pause on a sunflower in Iowa (opposite), and one takes off against the sun to continue the journey (top). The sun's position is a critical navigational tool in the monarch's migration across thousands of miles to a destination it has never seen. Entomologists believe the Earth's geomagnetic field may also have a role in orienting the butterflies.

On mountains in a limited area in the state of Michoacán,
they cluster tightly for the winter on their preferred oyamel trees.

An animal of extremes, the sperm whale roams a world of extremes. This enormous creature—the world's most formidable predator with the world's biggest brain—lives in three-dimensional space, its movements as much vertical as horizontal, its ocean wanderings taking it to the far corners of the globe. *Physeter macrocephalus* spends about three-quarters of its life in the blackness and grinding pressure of the deep ocean, sometimes diving almost two miles down to forage for food. As Herman Melville, who immortalized the sperm whale in *Moby-Dick,* wrote: "That head upon which the upper sun now gleams, has moved amid the earth's foundations."

PLUMBING THE DEPTHS

And yet, the planetary clock exerts a pull on the movements of this massive marine mammal as it does on even the smallest of creatures. Once a year, male sperm whales may cease their lone roaming and move toward warmer waters where groups of females and immatures live. Gliding out of the depths, the males announce their presence with a series of slow "clicks." Really more a clanging sound, these particular clicks are so loud that the females may be able to hear them from as far as 35 miles away.

The loudest sound made by any animal, the sperm whale's clicks emanate from its enormous nose, which takes up about a quarter of the length of its body and wraps around the wax-filled spermaceti organ. Some clicks form patterned codas that are clearly the means by which the whales communicate with one another. But, like so much about these animals, the slow clicks remain a mystery. Are the whales echolocating off other whales or boats? Are they warding off other males? Are they signaling their presence, or maybe their size, to the females?

Whatever the purpose of the clicks, the females surely hear them and can anticipate the males' arrival before their immense shapes looms shadow into view. As long as 60 feet and weighing in at as much as 50 tons, the males are almost a third larger in size than the females. Yet despite their Moby Dick reputations, male sperm whales are, in the company of females and young, the ultimate gentle giants. As the males join groups of females and immatures, they become "the focus of intense attention," cetologist Jonathan Gordon reports. "Calves, immature whales and mature females were all seen to push up against the mature males and roll along their backs."

The males, more and more solitary in their roaming as they age, apparently welcome the yearly attention, sometimes lightly gripping the young in their enormous jaws. Even the presence of other males

A SPERM WHALE is perfectly at home in the depths of the ocean. It spends about three-fourths of every day diving as far as two miles down to feed.

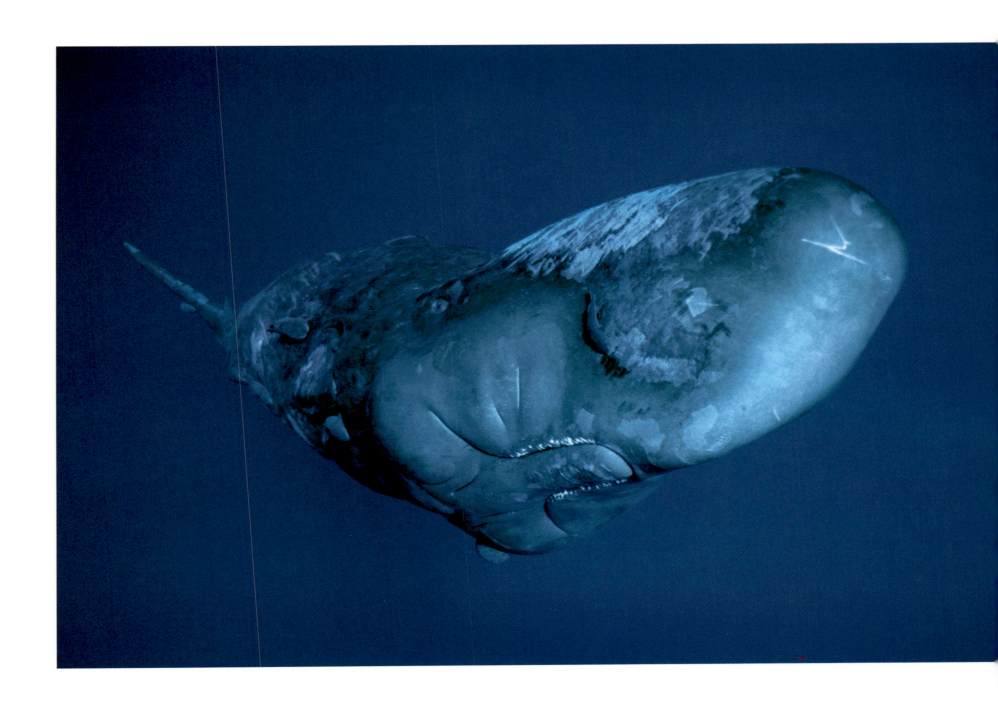

Pods of whales, often accompanied by their calves, find plentiful food in Caribbean waters off Dominica. A male (right) nuzzles a female during the breeding season.

on the breeding grounds doesn't seem to incite the normal male territoriality. Any sparring between competitors appears to be rare and over quickly, despite the grisly tales of old Yankee whalers. Instead of fighting on the breeding grounds, the younger males sometimes greet each other with a show of physical stroking that to the human eye looks affectionate or even sexual.

The kind of nuzzling and stroking that males experience on the breeding grounds is part of the daily fabric of life for females and younger whales. They live in a highly sociable world, traveling together in groups of 20 to 30 other animals related loosely through the female line. These "sisters," "cousins," and "aunts" live cooperatively, foraging for food together and sharing in the care of the young, even nursing each other's calves. While most of the group dive for food, a female or immature will often stay with the vulnerable young at the surface, guarding them.

In the past, whalers posed the greatest threats, taking sperm whales for their blubber and their coveted spermaceti. But with whaling now largely banned, the sperm whales' only serious enemy is the orca, or killer whale. Like the sperm whales, female orcas are social animals that travel in related pods, and their hunting skills are well honed. Orca attacks are so finely coordinated that they can take on far more massive whales, as they go after their young.

Sperm whales themselves are impressive predators, and collectively, they eat in a year as much tonnage as is taken by all the world's fisheries. For about 75 percent of every day, these toothed whales forage far below the sea surface, hunting particularly for the large squid that live in the depths. But for roughly a quarter of most days, a sperm whale group will gather together at the surface, nuzzling, touching, and generally socializing.

Female sperm whales reproduce slowly, giving birth once every four to twenty years. Each new calf is a critical link in the species' survival, and they take great care of the young through a long childhood. When males reach about ten years old, they leave the group, but the young females remain in the temperate ocean basin where they were born, part of a social group committed to each other and to raising the next generation of young.

The males may strike out for the cold waters of the Arctic or Antarctic, sometimes roaming in loosely connected bachelor groups. But as they age, they become solitary travelers, wandering thousands of miles a year alone. Over their long 50- to 60-year lives, they may tunnel a million miles through the Earth's oceans. But when the time is right, they move again toward the warmer waters and the females that sustain the species.

A SPERM WHALE CALF stays close to its mother (top left and above) and may not reach independence until it is ten years old. Females and their offspring gather at the surface after a foraging dive (opposite).

Exactly how many sperm whales inhabit the Earth's oceans remains unknown. Certainly there are many more than there were when commercial whaling ended in 1987. In the decades before that, mechanized ships concentrated their might against the large males, with their wealth of valuable spermaceti oil. With fewer mature males visiting the breeding grounds, sperm whale populations went into decline, and these massive deep divers were considered a vulnerable species. Now, several decades after the ban on whaling, sperm whales have made a slow comeback, and the technology to track them and better study them has made rapid advances.

Only in the past 30 years have researchers begun to study this elusive animal systematically and in its own habitat. A handful of committed cetologists have devoted their lives to tracking and understanding sperm whales as they move across the open oceans. Besides firsthand observations and photo identification, these scientists rely on radio-tagging to trace the animals' movements and on hydrophones to record their click communications. Even still, much of the sperm whales' world remains unknown. The good news, though, is that today, more than a million of them may be out there, moving, as Melville imagined, "amid the earth's foundations."

WHITE SPERM WHALES such as this one traveling with its mother (left) are extremely rare and perhaps are albinos. A sub-adult male dives spectacularly in cold northern waters (above).

NEED TO

"What a strange thing is the propagation of life!" the poet Lord Byron lamented. And yet the need to breed is the very backbone of existence. The breeding season can turn parents into warriors and competitors into killers. But in the end it is always the ultimate celebration of life. • That celebration can involve a true gathering of clans, as it does in the Falkland Islands every year, when **ELEPHANT SEALS, ALBATROSSES,** and **PENGUINS** return to reunite for the annual dance of procreation. In the north **PACIFIC SALMON** make the

BREED

long swim back to the Alaska brooks and rivers where they were born for their final act, a once-in-a-lifetime spawning before they die. Across the plains of East Africa, almost a million **WHITE-EARED KOB** gather on their breeding grounds, the males fighting brutally to defend their mating grounds. And in the forests of Costa Rica, ten million **ARMY ANTS** busy themselves day and night with the needs of queen and colony, their constant scurrying migration timed to the tempo of the queen's fertility and the collective need to breed.

Lying just beyond the polar maelstrom, the Falklands are circled by relentless ocean winds and currents. But every year, those winds and currents draw in an exuberance of life virtually unmatched on the planet. When the austral spring descends on these sub-Antarctic islands and the sun shines, their coasts become a riot of sound and motion, a celebration of breeding, birthing, and playing the odds for survival. The odds aren't really in their favor, but they come anyway—elephant seals, rockhopper penguins, and black-browed albatrosses—for the annual dance of perpetuation.

GATHERING OF THE CLANS

In September the first rockhopper penguins come ashore on Falkland beaches. Unerringly, they hop up into the rough cliffs above the sea, searching for one small patch of ground—the site of their nest from the year before. For half a year they have gone their own way in the Southern Ocean—the circumpolar waters surrounding the Antarctic. Now males and females reunite at the nest site; some have been together for years, others for life. As more and more rockhoppers return to the colony and to their nest sites, the quick staccato braying of thousands of birds fills the cliff face.

They don't have the cliffs to themselves. Black-browed albatrosses also crowd in by the thousands. And keeping a close watch on this seasonal reunion are the year-round residents of the Falklands—the striated caracaras, or Johnny Rooks as locals call them. They have waited through a long winter for the return of their prey and timed their own breeding to this annual reunion. They've even positioned their nests at the edge of the rockhopper/albatross colonies, waiting for

the eggs and chicks to appear. While they wait, they sometimes swoop down to the ocean's edge, to peck and pull at the skin of an elephant seal who has also come to find a mate.

The rockhoppers greet their returning mates effusively, the males swinging their crested heads from side to side and upward. They point their beaks to the sky in triumph and bray mating calls into the wind. Partners preen each other's heads and throats and deposit bits of stone or vegetation at the nest site. This is where they will mate, often copulating a number of times, the male standing on the prone female. And then they wait. About a week later, the female lays her first egg, but only about 60 percent of the first eggs produce a healthy chick. Then a second larger, more viable egg follows.

For the next roughly five weeks, the penguin couple will take turns incubating their brood. The first two weeks, they take turns sitting on the nest, relieving each other so both can feed in the sea. Then the female takes over the responsibility solo as the male goes to sea and

STRIATED CARACARAS, mighty members of the falcon family, nicknamed Johnny Rooks by Falkland Islanders, await the annual appearance of their favored prey, the chicks and eggs of rockhopper penguins.

ROCKHOPPER PENGUINS come ashore on the Falkland Islands to breed after spending many months away on the Southern Ocean.

To reach the highlands where they form huge nesting colonies,
they must climb steep cliffs—a remarkable feat for a bird that seems absolutely unfit for climbing.

forages for food. By the time the female's nest-sitting duty is done, she's ravenous and ready to return to the sea to replenish herself. For the final two weeks the male becomes the keeper of the egg, or eggs, if both have survived.

Weighing in at about six and a half pounds, the rockhoppers are among the smallest members of the penguin family, but they're fearless and ready to defend their ground against all comers. During the exchange of nest duties between male and female, the caracaras watch carefully, hoping to dance into the melee and grab an egg that becomes dislodged from the nest. Remarkably fast on their feet, the Johnny Rooks run with their prizes to their own nearby nests and present their hatchlings with a rich meal.

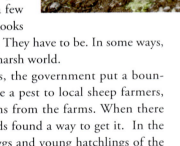

Members of the falcon family, these clever predators "partake of the form and nature both of the hawk and the crow," according to a Nantucket sealer who first visited the Falklands in 1812. He called them "flying devils" and "winged pirates." Found only here and on a few islands off Cape Horn, the Johnny Rooks are certainly clever and opportunist. They have to be. In some ways, they are the truest survivors in this harsh world.

In the decades before the 1970s, the government put a bounty on their heads because they were a pest to local sheep farmers, attacking lambs, even stealing items from the farms. When there was booty to be had, the pirate birds found a way to get it. In the Falkland spring, the booty is the eggs and young hatchlings of the rockhoppers and albatrosses.

ROCKHOPPERS (left) gather on the Falkland Islands as their breeding season begins. Meanwhile, rockhoppers on Saunders Island (above) are already tending the eggs of a new generation. Following pages: Rockhopper penguins spread out on their vast nesting grounds.

For the rockhoppers who've managed to bring their eggs to maturity, the hatching is a moment of triumph. Slowly, over the course of a day or two, the egg they've lavished so much care on cracks, and the chick appears. For two or three days, both parents remain near the nest, guarding it. The Johnny Rooks are still nearby, their yellow eyes scanning the thousands of birds in the cliff colony, hoping to spot an unattended or weak chick or even a weak adult.

A few days after a rockhopper egg hatches, the mother rockhopper begins to make short trips to the ocean to forage. Gradually, her trips become longer, as she dives for the small crustaceans—the krill—that will feed her brood. She returns in the afternoons and hops up into the cliffs to regurgitate her catch into the nest. The male meanwhile stands guard, never leaving sight of the hatchling.

After a few weeks, the young rockhopper is as big or bigger than its parents. Both mother and father now feel comfortable leaving the chick for a while as they forage in the sea. While the parents are away, the young penguins crowd together in crèches, instinctively knowing that there is safety in numbers. Once the chicks molt into their adult feathers, they are ready to flex their wings, learning to fly through the cold, food-rich ocean.

While the birds occupy the high ground in the Falklands during the annual breeding season, elephant seals claim the surf's edge. The first bulls begin to arrive in September, their great carcasses—as much as four tons and 15 feet long— hauled out on the beaches as they wait for the females to arrive. The females follow in a few weeks, always returning to the same beach as the season before. But until they haul out to mate, they continue to forage. Diving a thousand, two thousand feet, they fatten up on squid and the fish in the depths. Once they come ashore, there'll be no time for feeding. They'll be bombarded by demands on them and their energy from all sides.

Meanwhile, the first male arrivals are using up energy fast, defending their beach territory as more and more males haul out. The early arrivers have made a calculated error: By the time the cows do show up, these males will be exhausted, too tired to defend territory or females when the mating games begin. They will be consigned to the edges of the harem crowd.

The cows are far more congenial than the males, mounding in crowded clusters on the beach, gregarious and easy with one another, but never truly touching—an inadvertent touch can elicit a snarl. More and more females haul out onto the beach, until several thousand seals color the shoreline like enormous rocks.

While the males bellow, challenge, and charge each other, establishing their claim to be a beachmaster, the females have their own

A FOUR-TON SOUTHERN ELEPHANT SEAL (opposite) rests in the surf after arriving in the Falklands to breed. The youngsters receive careful attention from their mothers (top right). The females are not always treated kindly by the herd's bulls (above left).

urgent task. All year they've been carrying a calf, conceived the previous year on this beach. Once on dry land, they give birth to pups covered in woolly black down. But the tumultuous, crowded Falkland beach is a dangerous place for the newborn seals. Some 20 to 30 percent of the pups won't survive, many of them trampled in the fracas that ensues as the stakes increase.

For the first three weeks, the mother seal spends a lot of her time nursing, fortifying the pup with milk so rich that the fat laid on now will keep the pup nourished for at least the next two months, as it learns to forage on its own. Then, about 20 days after giving birth, the cow goes into heat, and her focus moves from maternal care to mating.

The bulls are on the lookout, and one soon approaches—sometimes the alpha beachmaster, sometimes another bull. Thousands of pounds of sexual aggression, the bull grabs her neck with his teeth as they couple. Her whimpering may alert the beachmaster, if he is not her partner, and if he is not already engaged himself. Charging over to reassert his claim to the female, he roars through the enormous proboscis—trunk of a nose—that earned these animals' their name. That may be enough to run his competitor off. If not, the bulls are ready to fight. They maneuver around each other like sumo wrestlers. They thwack necks and bare their teeth. Blood from puncture wounds soon darkens their shoulders and necks, until finally a victor emerges—usually the beachmaster. The role of alpha male is

AFTER ARRIVING ON THE BEACH, elephant seal bulls (above) are quickly ready to begin fighting for territories and mates. The bellowing bulls—weighing up to four tons and as much as 15 feet long— are doing more than boasting (right). Their battles are often vicious, and opponents may be severely injured. The top winner becomes the colony's "beachmaster."

too exhausting to last for many seasons, but during even one season, a beachmaster can pass his genes on to as many as 80 females, while unlucky males may never mate.

Once the mating ritual is completed, the impregnated cows leave their pups and the Falklands behind. Emaciated and desperate to feed, they return to the deep ocean. The exhausted beachmaster will follow soon. But next year when the austral spring softens the circumpolar air, they will all convene again on the same beach.

The seals and rockhoppers move through their breeding and birthing cycles faster than the black-browed albatrosses, who share the vast colonies on the Falklands' coastal cliffs with the penguins. These islands are one of the few places where the magnificent ocean-wandering albatrosses make landfall, swooping in for an ungainly, almost head-over-heels landing every spring. They are the ultimate flying machine, their bodies made to catch the ocean currents and soar dynamically along with barely a wing beat.

"Alcatrazes is againe the biggest of any Seaffowle I have yet seene," wrote sailor Peter Mundy in 1638, "spreading Near 6 or 7 Foote with his wings, which hee seemeth not to Move at all as hee Flyeth leaisurely and close to the Rymme of the water."

Though the wings of the albatross are amazingly long, they're also narrow, a design that perfectly exploits the continual winds along the sea surface. With their seesawing flight, they can reach speeds of over 60 miles an hour with little expenditure of energy. It's the takeoff and landing that costs them real exertion.

The mollymawks, or black-browed albatrosses, are one of the smallest and most numerous members of the albatross family, and more than half of them choose the Falklands as their breeding base. By early September they have begun to arrive. Like the rockhoppers, the albatrosses reunite with their mates on the same nests they occupied in previous years.

More elaborate than the rockhoppers' simple ones, the columnar albatross nests are built of peat or mud that the birds excavate. From year to year the albatross couple refurbishes or repairs their nest. Also like the rockhoppers, the albatrosses mate on their nests, but it takes six weeks before the one precious egg is laid. During that time, the male stays close by the nest, while the female continues her ocean wandering as the egg she's carrying ripens. Her search for food can easily take her several hundred miles beyond the Falklands.

When the time comes for laying, the female returns to the nest. It will take another ten weeks of careful incubation for the egg to hatch, and both parents take turns patiently sitting on the nest.

AFTER WINTERING as far from the Falklands as South Africa, black-browed albatrosses form a colony of their own, and birds of a breeding pair groom each other's neck feathers (opposite). A pair's single chick may not survive; Johnny Rook falcons drop in for a kill (top left), and many chicks become a meal (above right).

Even the hatching process takes time—up to four days. At last the chick emerges, and the parents become even more vigilant, knowing that the Johnny Rooks are still watching, always prepared to run the colony gauntlet to grab the unsuspecting young.

For another three weeks, one or both albatross parents are on hand to tend the chick. Then they take to their true milieu, the air. Normally, they circle north, feeding along the coasts of South America and returning to the nest and the chick periodically with food. It has been six months since the couple first reunited in the Falklands, and the wind is beginning to carry hints of the fall—a sign that the time has come for the black-browed fledglings to become airborne. When they do at last take off, it will be for the long haul. For the next ten years, they'll soar on wind currents that will carry them around the Southern Ocean and into the latitudes off South America and South Africa.

By winter, the colony cliffs are still and deserted, scattered with the corpses of the young ones who didn't make it or who fell prey to Johnny Rooks. The caracaras' own young are now fully fledged and honing their survival skills. They wander through the deserted colonies, hoping to scavenge something left behind. Many won't make it through the winter months ahead.

And many of the seasonal sojourners won't make it back to the Falklands. For the rockhoppers, the great danger used to be from sealers, who took the penguins for their fat and their coats. That danger ended almost a century ago, but rockhopper numbers have fallen precipitously recently, down 30 percent over as many years. Oil spills take a toll, and commercial fisheries now compete with them for the food they need to survive, but there may be other, unknown reasons for their demise.

The albatross's fate seems strangely foretold in Samuel Taylor Coleridge's *Rime of the Ancient Mariner.* In it the albatross ". . . ate the food it ne'er had eat / And round and round it flew . . . / And every day, for food or play, / Came to the mariner's hollo." Ultimately, the mariner shot the albatross.

Today's albatrosses could be doomed by modern mariners. The mollymawks have been hit hard, if not literally by the mariners themselves, then by their methods. The birds opportunistically follow behind longline fishing boats and trawlers, feeding on their leavings. When they gulp bait on the longline hooks, they're often snagged and drowned.

There are recent positive signs, though, that the albatrosses are making a recovery. In coming decades, maybe not just thousands but the millions of birds of the past will return to the Falklands each spring, and the cycle of breeding, birth, predation, and survival will be assured.

MUCH LARGER BIRDS than black-browed albatrosses are wandering albatrosses (opposite), here performing their courtship ritual. A black-browed albatross (above left and right), in its first few moments of flight, takes leave of the security of land for the prospects of food at sea.

ALBATROSSES ARE EXPERT aerialists in the Southern Ocean's gale-force winds, their long wings adapted with exquisite aerodynamic design to master the weather.

A light-mantled sooty albatross soars over South Georgia Island,
where the species nests in open areas away from the top of high cliffs.

Army ants move to an almost unfathomable instinct, their seething colonies of 500,000 to two million individuals so finely tuned that they operate as if they're the cells of a single organism. They are almost all female, part of a related sisterhood, and as with every other animal, they are consumed by the same drive—to keep the young alive and flourishing. That drive makes them formidable predators, devouring virtually anything in their path. Like any army, they ravage the territory around their camp. And then they move on.

A MOVABLE FEAST

As the first streams of sunlight hit the ants' bivouac deep in the Central American rain forest, the colony begins to stir. Late the night before, around midnight, the workers linked their claws together to create chains and clusters with their bodies. Now, though unsighted, the ants can sense the dawn light, and they begin to disentangle and tend to the business at hand—a day of anxious foraging for food to feed thousands but particularly the immature members of the colony.

The vanguard of workers, the small and medium-size members, move out first, releasing a trail of pheromones from their gasters, laying the trail back and forth for the others to follow. Fanning out behind them, the rest of the army ant colony scrambles forward, on the hunt, anxious to find food. But no one strays very far away from the swarm—the ants seem, as entomologist William Gotwald explains it, "shackled to their chemical trails, which they appear to follow slavishly."

Key to each colony of army ants is the queen, bulbous and regally larger than her sterile subjects. While they work, she keeps to the nest, periodically and on schedule tending to her primary job—laying eggs,

thousands of them. Now, as the swarm starts its dawn-to-dusk foraging, the queen stays in the bivouac with the developing larvae.

On the edges of the army, the soldiers, equipped with hooked mandibles and pincers, guard the flank. The forest literally seethes as the swarm advances and other insects, even small animals, rush to get out of its path. Its camp followers, antbirds and antwrens, herald the approaching army, though the ants, deaf and blind, are oblivious to the sound. The birds are opportunists, feeding on the insects and other arthropods flushed by the advancing army. And following along in the wake of this motley legion are butterflies that feast on the birds' droppings.

The ants have no mercy. An eerie ellipse as much as 45 feet wide, each column marches on, sometimes swinging left or right but maintaining form and function. Some of the columns sweep the leaf litter, others march up into the trees. When they encounter chasms and obstacles, they again instinctively join their bodies together to build living bridges for those behind them to cross, so that the raid is never stopped or slowed.

AN ARMY ANT is literally armed to the teeth, its sharp-pointed jaws gigantically oversize in comparison with the head. Imagine the havoc a colony of two million can cause.

Workers with huge mandibles guard the entrance to a temporary cache where the ants store
prey captured during a raid before carrying it back to the nest.

Tarantulas, scorpions, grasshoppers, wasps, even reptiles and small birds are no match for the swarm. It hunts by smell and by sensing motion, intuitively scurrying toward anything on the move. The foraging workers attack with precision, stinging or asphyxiating their prey, even if the prey is a hundred times their size. The soldiers move in to dismember the catch with their mandibles, bladed with a row of sharp teeth. Once the quarry is in manageable pieces, the long-legged porters take up the colony's burden, transporting the dismembered meal to the bivouac. During an average raid, they may carry as much as 30,000 bits of prey back to the camp.

Even when dusk seeps into the forest, the ants' work isn't done. Before their day ends, they have to transport the larvae to a new bivouac site about a hundred yards down the trail laid earlier that day. When that is accomplished, the queen emerges, surrounded by her workers and soldiers. By midnight she is ensconced with her colony in the new bivouac, and the workers and soldiers have linked their claws, forming protective nets with their bodies.

The Germans call army ants *Wandermeisen,* but their wanderings are done in carefully regulated cycles. For two to three weeks they are migratory, on the move to a new bivouac each night. Then for about three weeks, they are more or less stationary, staying in the same bivouac and launching foraging raids out from it during the day. Unlike most animals, their migrations have nothing to do with seasonal signals—the movement of the sun, wet and dry seasons—or any other external force. The colony apparently creates its own migratory tempo, and that tempo has something to do with the reproductive phases of the queen and the stages of development of the young.

As each nomadic phase draws to an end, the larvae stop their manic feeding and begin to spin cocoons. The workers, so busy foraging and bringing food back to the bivouac and moving the colony each night, continue their programmed behavior, but now there is a surfeit of food, which the queen feasts on. She adds fat to her body, and the fertilized eggs she is carrying begin to mature more quickly.

It's time for the wanderings to end, and the colony to go to ground. Once again the workers shoulder their burdens, this time the cocoons almost ready for pupation, and head to a semipermanent bivouac they have used in the past, sometimes inside a hollow log. The queen and her retinue follow the hormone-laced trail to the new site. After a week or so, she begins laying the new batch of eggs, some 100,000 to 300,000 of them.

RAPACIOUS HORDES OF ARMY ANTS (opposite) blanket the layer of leaf litter in a Central American rain forest. Successful attackers feed on a grasshopper (above).

In a sudden burst of fecundity, the eggs hatch into larvae. A few days later the cocoons too pupate, and new "callow workers" emerge. At first immature and weak in both physical strength and coloration, they are still recognized by their "aunts" as the new crop of workers that will help keep the colony alive.

For as long as three weeks, the colony has been in one place, with workers, soldiers, and porters foraging out from it every day to find food to bring back to the bivouac. By now they have devastated the available supply in the surrounding forest. With the new callow workers viable, it's time to move on to greener pastures, or at least to an area of the forest where the prey population hasn't been depleted by weeks of colony feeding.

The colony begins a frenzy of activity as the workers prepare once again for the migratory phase. Foraging by day, the workers move the brood of larvae and the queen to a new bivouac site each night. Then, after a few brief hours of rest, they're on the move again, sweeping the forest for food. Once a year, early on in the dry season, the normal reproductive cycle changes and the chronic egg-laying and larval-development tempo is broken as a new note is introduced to colony life. The queen gets ready to lay the annual sexual brood. Once a year, she produces a vital contingent of males from unfertilized eggs and several new queens from fertilized eggs.

When the new queens emerge from their cocoons, the workers huddle around them. A few days later, the males pupate. Things in the colony are about to change dramatically. As the migration begins anew, the workers split, creating two distinct pheromone-scented trails from the bivouac site. The new queens and an entourage of workers will take one, while the old queen will take another. The colony has split in two, but despite what seems like a fait accompli, the fate of the queens is not yet decided. Only one of the young queens makes it to the new bivouac.

The old queen will continue to reign supreme with the remainder of her now-bifurcated hive until she begins to lose her charm or effectiveness. She's the glue that holds the colony together, and if her age is affecting her power to do that, her subjects have no mercy. She, too, will be held back from the new bivouac and a virgin queen allowed to take her place.

The males are also cast out by their aunts, perhaps as an instinctive way to keep the gene pool healthy. Unlike their female kin, the males have eyes and can fly. They have only a week or two to live and only one thing they have been designed for—mating with a queen. They are literally, according to social biologist E. O. Wilson, "flying sperm dispensers." In order to accomplish their life's task, the males have to find a new swarm outside their birth colony. And once they've

ARMY ANTS PATROL a trail in the tropical forest (above left and right). When the ants raid a killer bee colony, the bees are no match for the marauders (opposite). Following pages: Army ants groom a larger ant (left) and, locking claws, form a temporary bivouac (right).

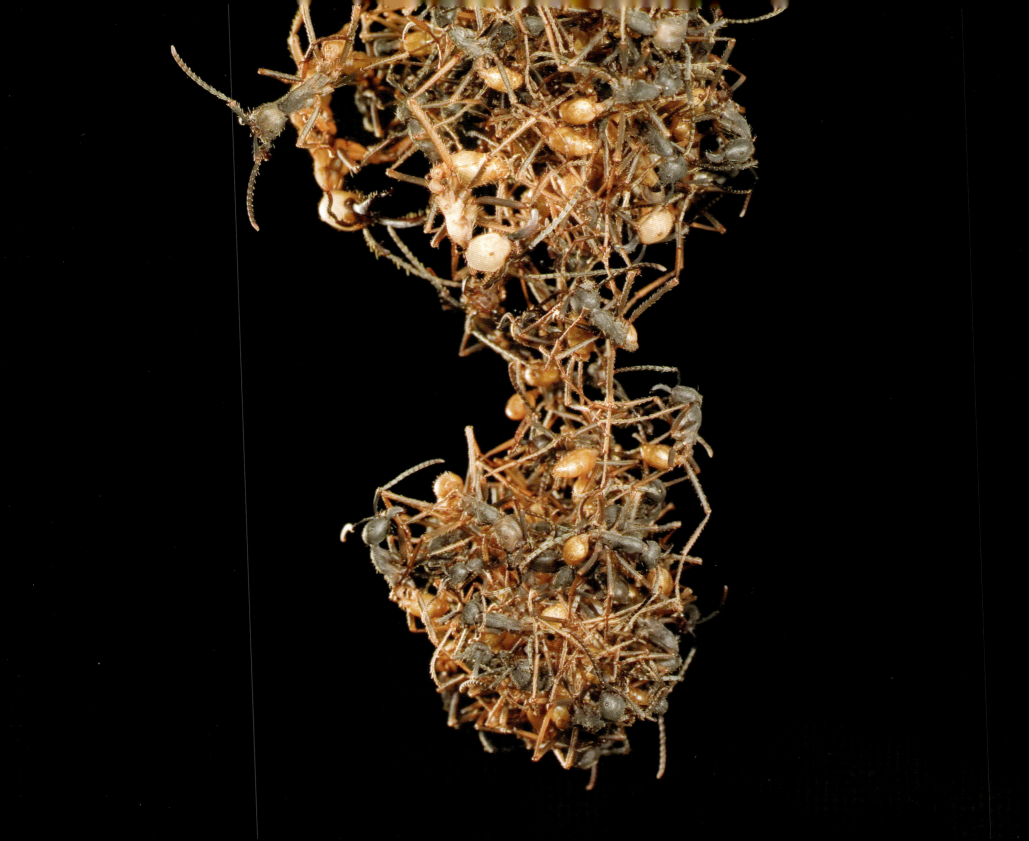

found it, they have to get beyond the gatekeepers, the workers that surround the queen. They accomplish that by being large and energetic enough to impress the workers. Instinctively, the colony selects the "best" males to create the colony's progeny. Only a few males are allowed to pass on their genes.

As always, the colony acts with one mind, regulated by collective need rather than the needs of any individual. Everyone has a job to do to make the organism of the colony function, from care of the young to foraging for food. This highly evolved social system goes back deep into evolutionary history, to the Cretaceous period, far earlier than human fossils have been discovered.

Unlike the ants, "we haven't evolved in the societies we currently live in," animal behaviorist Iain Couzin believes. He and other scientists are studying the cooperative behavior of the swarm to understand better what makes it tick. Couzin wants to understand "how the type of feedbacks in these groups is like the ones in the brain that allows humans to make decisions."

"These marvelous little creatures have been on Earth for more than 140 million years," says biologist Wilson, a great admirer of ants. He ranks the complex social organization of army ants as one of "Earth's greatest wildlife spectacles," saying "Ants easily outlasted the dinosaurs, and they will easily outlast humanity should we stumble."

When dawn lightens the forest floor, the ants will be on the move again, an advancing army programmed chemically to attack, feed, defend, and perpetuate the vast organism that gives them purpose and life—the colony.

NOMADIC ARMY ANTS migrate to a new location at night (above). The workers' duties include carrying the colony's pre-adult pupae while migrating (right).

In the aftermath of war, you expect to find devastation. You don't expect to find a natural grandeur long believed to be lost. You don't expect to rediscover one of the greatest migratory spectacles on Earth. But there it is, some 1.3 million white-eared kob, tiang, and mongalla gazelle, filing across a landscape where just a few years before, war raged, seeming to suck the life out of the boundless savannas. Yet the urge to create life was stronger and more relentless, winning out in the end against war and destruction.

ALL'S FAIR IN LOVE AND WAR

"I have never seen wildlife in such numbers, not even when flying over the mass migrations of the Serengeti," says J. Michael Fay. In 2007 Fay and Paul Elkan, with the Wildlife Conservation Society, flew across this war-plagued corner of the Earth, conducting an aerial survey of wildlife in Southern Sudan's national parks. After 15 years of fighting here, fighting that had ended only a couple of years before, the scientists didn't know what they'd find. It had been 30 years since the last aerial survey of the area. But after 150 hours surveying some 58,000 square miles of Boma and Southern Sudan parks, they knew they'd struck it rich beyond their wildest dreams. It was, Fay said, "like discovering the wildlife version of the *Titanic*."

They had counted some 800,000 white-eared kob, an antelope believed to be in decline, as well as tiang (African antelope), and mongalla gazelle. In subsequent flights, they spotted long-horned oryx, thought to be extinct in Southern Sudan. And they counted nearly 4,000 endangered Nile lechwe—an antelope found only in this region—grazing in the Sudd. In the woodlands to the east of the Sudd, the survey team also saw the tracks of thousands of elephants.

One of the world's largest wetlands, the Sudd expands with the seasonal flooding of the White Nile. A sodden natural barrier to Southern Sudan, it has kept invaders, poachers, and almost any outsiders from encroaching on the area. For centuries the Dinka, Murle, and other indigenous people grazed their cattle on the rich grasslands of the Sudd, but that ended in the early 1980s, when their cattle and villages were decimated and civil war made perpetual refugees of many of them. Somehow the migratory animals—the antelope and gazelles and oryx—stayed ahead of the fighting and apparently flourished in spite of the chaos.

The white-eared kob dominate here, hundreds of thousands of them filing through the savanna, their paths, according to Fay, "like the paths of driver ants through the dense grass, all moving at a frantic pace north." Like the wildebeests of the Serengeti, these antelopes are almost perpetually on the move, covering hundreds of miles in a year. By November the relentless sun of the equatorial dry season has turned the water sources in their southern breeding grounds to mud, mud that becomes a sucking trap for their young. It's time to move.

THE REGAL WHITE-EARED KOB, an antelope thought to be in decline, turned up in unimagined numbers—800,000—during an aerial survey of Southern Sudan's wildlife in 2007.

HUGE HERDS of white-eared kob cross the plains of Southern Sudan, tracking the moisture and grasses of the rainy season.

All too soon, the grasses begin to dry up, and the herds must follow
the rains to new pastures on the savanna.

With wildfires chasing them and their own hormonal cues pushing them forward, they head toward their breeding grounds on the east side of the White Nile, near the Sudanese-Ethiopian border. This is the Boma-Jonglei region, an area about the size of New York and dominated by East Africa's largest, most intact savanna ecosystem.

Boma National Park is crisscrossed by a network of rivers and swamps that leach moisture into the meadows where the kob graze. The nutrient-rich "black cotton" soil here is tufted with savanna grasses that continually leaf out in new, protein-laden sprouts. In the heat of the day the kob take refuge in the shade of acacia woodlands. But at night they're back on the meadows, grazing.

Scattered across the savanna, the animals seem to be foraging at random, their placement haphazard. But here on the summer breeding grounds, they're actually clustering themselves into self-selected groups, the dominant males marking off circular territories for their leks, or breeding arenas. They warn off other males by lowering their heads and feinting with their horns. If that doesn't work, serious fighting can ensue. The dueling males stand with their feet apart like prize-fighters, as they literally lock horns or crash against one another. The gored bodies of vanquished contenders bloody the ground, ground

that the winner holds at his continued peril. He may reign as champion of the lek for less than a day, or for over a month. And during that time, he has more to do than just ward off other males. The real point of the battle is to attract females in estrus.

The bull's strength and prowess in holding his territory is one way to do that—it signals to the females that he has the favored genes of a survivor. A bull may also try a more charming tactic to woo a mate, almost coquettishly prancing to win a female's attention and favors. But perfume is his greatest weapon—the scent of his urine tells her all kinds of complex information about his genetic possibilities, whether he is a good match for her, whether he carries diseases or parasites.

If she likes what she smells, she'll return the favor, and it will be his turn to decode the scent, screwing his face into a clownish expression as he analyzes it through special receptors in the roof of his mouth. If he feels she is receptive to breeding, the male will touch her underside with his foreleg, then mount her in a brief coupling. The mating concluded, their quixotic relationship is over and both move on to other partners.

In late March or April, the winds sweeping across the savanna shift from northerly to southerly and bring in wet ocean air from both the

A WHITE-EARED KOB faces frequent challenges. One challenge (top) is for dominance on the breeding grounds. Another, sometimes unsuccessful (opposite), is to survive an attack by cheetahs.

Atlantic and Indian Oceans. Gradually, the kob make a slow turn to the south. By July they've again reached their wet-season range and calving grounds, where there's less chance of flooding than in the north and plenty of forage.

Life on the move across this savanna world is good for the white-eared kob now. But there are signs of problems to come. As the refugees from the civil war return to their homeland, the kob's wide-open world will narrow. Already roads are being built with little thought to the migratory wandering of these antelopes and other ungulates; poachers armed with automatic weapons take full advantage of the new thoroughfares, using them to kill and transport game for the illegal bushmeat trade. Oil exploration, too, is increasing in Southern Sudan, never a good thing for wildlife.

While peace and the increased human presence have had some negative effects on the white-eared kob and their fellow migrants, there are also positive signs. The Wildlife Conservation Society is continuing its research on the kob and on ways to protect them. The society is currently working with the government to establish a new national park on the Bandingalo plains, one of the most intact savanna habitats tracts in Africa and a seasonal home to large mammals and birds. The area would also provide a critical link between Boma National Park and the Jonglei area, both on the kob's migratory circuit. Most important to these animals and all others in Southern Sudan is the growing public awareness there that parks and the wild creatures they protect are a valuable asset both to the country and to the natural heritage of the planet.

UGANDA KOB PAUSE during migration to graze on the dry savanna in Virunga National Park (right). White-eared kob are quick to sense the approach of a potential predator (above).

Alaska summers arrive like brief blessings. Long days of sun pull life out of the dormancy of winter. The frigid months of survival become a summer frenzy to survive. What happens now—what gets eaten, stored away, hatched, or bred—will make the difference between life and death. Every animal instinctively knows that. They range through the forests, coasts, and rivers, searching for food or others of their own kind. Some, like the many kinds of Pacific salmon, make the pilgrimage back to the place of their birth to spawn. It's their final task and accomplishing it will be the death of them, one way or another.

THE SEASON OF SURVIVAL

Water temperature and hormones signal the salmon that it's time to make their way home. Some have been circling the Pacific for a year, but most for two or three or five years. Now magnetite particles in their brains, sensitive to the Earth's geomagnetic field, and polarized sunlight filtering through the ocean waters will guide them back to the place of their birth. They bring with them the vital nutrients of the ocean that they've ingested. Their bodies will eventually serve as fertilizer, feeding the forests and the fauna of the whole Alaska ecosystem.

Knowing that the salmon are beginning the long swim back to Alaska waters, their predators begin to assemble. The first line in the predatory gauntlet are the salmon sharks, fish so large and ferocious that they're sometimes mistaken for great whites. Sensing the patrolling sharks, the returning salmon make the mistake of hugging the shoreline, inadvertently trapping themselves. The salmon sharks move in for the kill. But the sharks aren't the only predators hoping to gorge on salmon. Sea lions and whales are gathered in the sound as

well, hunting the returning chinook, chum, coho, pink, and sockeye salmon. Mostly the predators win out. Only one of every four salmon survives to leave the sound.

The instinct to breed drives them back to their birthplace, their spawning site, and their sense of smell guides them. For some, the spawning grounds aren't far from the salt waters of Alaska's filigreed coastline. Others have a long road ahead to reach the inland lake, river, and stream of their birth.

Bald eagles, gulls, river otters, minks, and Alaska's most formidable predators—brown bears—are all eyeing the watercourses, waiting for the annual salmon run. But the predatory gauntlet is only part of the challenge. The inland-breeding salmon will be swimming upstream, against the current. When they hit obstacles—rapids, shoals, waterfalls—their determined biological drive keeps them going. They'll literally take a leap of faith, hoping to fling themselves beyond whatever heights they find in their paths.

SOCKEYE SALMON are in flamboyant breeding attire when they return to their ancestral spawning grounds, which for this trio is the Horsefly River in British Columbia.

DURING THE SPAWNING SEASON, hundreds of pink salmon of all ages pack the rivers in Alaska, where each new generation grows up in the same place the parents were born.

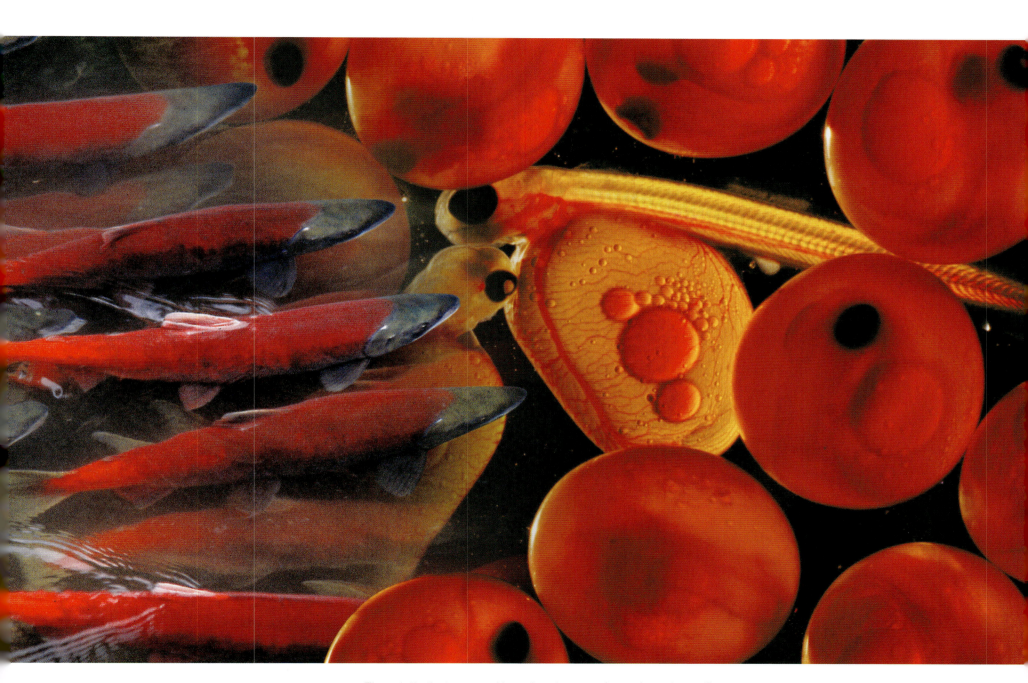

The adults that succeed in swimming up a hazardous river will soon spawn,
and millions of tiny eggs will fill the water. Only a small proportion will hatch and reach adulthood.

They haven't fed for a while when they reach Alaska and are surviving on the fats they stored during their ocean foraging. At sea, both sexes glimmer silver in the ocean currents. But as their hormones changed, signaling the time to return home and breed, their bodies changed as well. The male's back began to get narrower and more humped, and his color began to intensify and change hue. Now, as he nears the spawning grounds he's become almost tropical in his flamboyance, sporting a green head and a crimson body—most extravagantly in the sockeye. The female too has reddened, and the salmon become a crimson tide coursing upstream, the males in the lead. If they make it to their destination, they'll stake out their territory and wait for the females to arrive.

But all along their route, the bears are waiting, driven by their own appetites. They've spent the winter in their dens, sleeping and dormant. They're lean themselves now and driven by the need to feed. In the summer sun days that never end, the bears devote their time to eating, eating as much as they can as fast as they can. Otherwise, they won't make it through the next winter.

Usually unsociable animals, they find themselves thrown together, collecting at the edges of the salmon streams. Watching the mature bears, the cubs, born in the winter dens, learn how to fish and interact with their own volatile species. Power and strength dominate, and the biggest, most powerful males have their way, virtually uncontested. They claim the best fishing holes on the streams and position themselves there—until a bigger, more powerful male muscles them out of the way.

All the bears covet the spots beside the waterfalls, chutes, and shoals, where the salmon pool up, readying themselves to become airborne in a leap that, if they make it, will propel them forward up the stream. And when they launch themselves skyward out of the white water, the brown bears are ready, grabbing them out of the air with their massive paws.

In the first flush of the feast, the bears eat the whole fish, desperate to replenish their bodies after the winter. But as hunger abates and the salmon run goes on, the bears become choosier, eating only the salmon's nutrition-laden brains, eggs, and skin. The carcasses they leave behind are scavenged by wolves, birds, and other animals. Whatever is left of the fish bodies dissolves into the soil, the chemicals finding their way into the forest canopy.

The salmon survivors, their numbers depleted at every streambed obstacle in their upstream path, keep going until they reach the gravel

FIGHTING UP AND OVER WATERFALLS on mountain rivers, salmon often make a meal for a massive predator. Alaska brown bears are expert fishers. Stalking and reaching for a fish amid the falls (above), they make a percentage of successful catches (opposite) that a human angler would envy.

beds they've been searching for. Once there, each female begins digging a depression called a redd in the gravel, a process that can take her as long as a week. Lying on her side, she fans her fin back and forth to move the gravel. Once she's excavated the site to her satisfaction, she lays her eggs in the redd. Now the male moves in, depositing his milt, or sperm, on top of the eggs. In her last act, the female again uses her body to displace the gravel on the streambed, gently swaying back and forth and fanning her fin to move gravel back on top of her fertilized eggs.

If she manages this, her swan song is successful. With the eggs secure, the female soon dies, her body adding the Pacific's nitrogen, phosphorous, carbon, and other vital nutrients to the streambed. The males, too, have completed their life cycle once spawning is over. The next generation has been set in motion, but the maturing embryos will be left to cope on their own. Only a few from each redd will survive.

For two to three months the clutch of pink eggs, each about the size of a green pea, develops. Then the hatchling emerges. Staying in the redd for the first month, the one-inch alevin gets its sustenance from the egg's yolk, still dangling from its body. Then it's ready to leave the nest as a tiny fry. Through the cold months ahead, it will stay in the frigid waters of its birthplace.

The following summer it begins its own long, downstream odyssey to the ocean. Along the way it will feed on plants and insect nymphs, and in turn be fed upon by birds, insects, and other fish. As it heads toward the sea, it will pass the returning generation of adult salmon moving upstream, coming home to spawn and die.

ATLANTIC SALMON, closely related to the Pacific species, release their eggs and milt (right). The recently hatched young, called alevin at this growth stage, cluster in large schools (above). Following pages: The iconic salmon migration nourishes the Alaskan landscape.

They glide through the tropical forest canopy, their arms and elongated fingers webbed and outstretched to form wings that can span an impressive five feet across. Despite their flying prowess, they're not birds, and, despite their name, they're not foxes. But they are flying mammals, one of the largest fruit bats in the world, their inquisitive, big-eyed faces and furred ears and muzzles earning them the name flying foxes. The little red flying foxes of Australia are truly aerial nomads, breeding and bearing their young in the treetops and following the trail of nectar and flowers wherever it leads them.

LIFE ON THE WING

These winged mammals are a marvel of endurance and engineering, survival smarts, and selective sociability. Their "camps" are small cities with thousands, even a million, animals clustered body to body in the treetops by day, hanging upside down as they sleep through the sunlit hours.

As dusk settles, the camp begins to stir, the winged arms of the bats unfold, and one by one by one the members of the colony lift into the air, as if a strong wind had blown thousands of leaves skyward from the treetops. The screeching, airborne flock sets off for a night of foraging that can be an endurance test, with the little reds flying as far as 50 miles in a night.

Unlike most bats, this species doesn't use echolocation to navigate the skies. Instead they follow their noses and use their keen, color-sensitive eyes to find food. Despite their reputation for ravaging fruit orchards, the little reds much prefer the blossoms and particularly the nectar of flowering trees, using their long tongue to sip from flowers. Their favorite food is calcium-rich eucalyptus blooms, but if the eucalyptus are in short supply they'll feast on shoots, bark, sap, other hardwood blossoms, insects, and fruit, which is what gave them such a bad reputation with orchardists in the past. The little reds will venture farther inland in their foraging than other bats, and as they move from flower to flower, they are pollinating, playing a critical role in regenerating the hardwood forests.

Returning to their camps as sunrise lightens the sky, the little reds swoop in low to the ground, where the wind resistance is least, and the danger from an unlikely source is greatest. Grabbed from the air not by predators but by the barbed wire fences of beef and dairy farms, they become ensnared. As they try to chew themselves free, they just exacerbate their injuries and become more tangled. Unless they're helped by a human hand, they won't make it out alive.

Most of the colony will make it back to camp, snatching at the limbs of roost trees, hooking themselves with the thumbs of their

AUSTRALIA'S LITTLE RED flying foxes are fruit bats, true aerialists whose winged arms and keen noses take them unfailingly on a quest for sweet blossoms and nectar.

Before night falls, they will form vast flocks and make a
nightly migration to fruit-filled forests, where they will feed.

wing-arms as they land. Then they fall into a hanging position amid much commotion and screeching. The little reds' landings aren't always smooth, and quarters are close in the camps, so territoriality can erupt into snapping, screaming encounters.

For most of the year males and females roost together. From November to January during the late spring mating season, males mark off territories and defend the several females they've attracted. Courtships flare into touching scenes of nuzzling and stroking. The pairings can take as long as 20 minutes, and afterward the couples, together for at least this season, sometimes continue to cling to each other for hours.

In April or May, after five months of gestation, the females will give birth, usually to one infant. For the first month or so, the new mothers will take their suckling babies along when they forage, the infant clinging to its mother's fur with its feet. Carrying this added load, the mother needs extra energy to sustain her. Childhood is brief for a little red, and by two months old, the young bats are ready to

fly on their own and follow the camp in its nomadic wandering.

Soon enough the nightly foraging depletes food supplies in the region around the camp, and within a month or two, it's time to move on again. Before breaking camp, the little reds carefully groom themselves, tonguing an oily secretion over their wing-arms that will make their flying more energy-efficient.

The treetops they leave behind in their abandoned camps are virtually denuded from their stay, as the takeoffs, landings, and roosting of the little reds have dislodged leaves and left guano-laden branches. Some branches have given way altogether under the weight of so many clustered bodies.

Though their trail today is obvious, their path through the deep past remains something of a mystery. The oldest fossil remains of any kind of bat dates to the Eocene epoch, some 50 million years ago—fairly recent in evolutionary terms. Yet today the order Chiroptera is one of the most diverse group of mammals, with more than a thousand species of bats.

Bats probably evolved from arboreal gliders, with a stretched membrane that allowed them to simulate flight. That membrane eventually extended from the hind limbs across the arms and the extremely elongated fingers. Only a clawed thumb was left sticking up from this all-encompassing wing. The little reds use their thumbs when they climb or grip fruit.

Highly adaptable creatures, little reds live wherever they find food, ranging across northern Australia and far down the southeast coast. Towns and cities, with their well-tended trees and watering sites,

A GRAY-HEADED FLYING FOX PUP hugs its mother tightly, her "thumbs" anchored to branches (opposite). Awake and big eyed in darkness (above right), or gorging on fruit (above left), flying foxes congregate in "camps" of thousands and move en masse in search of food. Following pages: A lone flying fox takes flight.

make good roost sites, and large colonies of these bats can be found in the suburbs of major cities like Sydney, Melbourne, and Brisbane. But colonies of the nomadic bats can also be found in the wild.

In a year's time, the little reds may cover a couple thousand miles in their wanderings, setting up their temporary camps in bamboo thickets, eucalyptus stands, paperbark swamps, and mangrove marshes. The mangroves particularly protect them from human predators. Though hunting them less than in the past, Aborigines still take occasional fruit bats for food. And when the little reds fly low over rivers, swooping down for a drink, Australia's infamous predator is sometimes waiting. Some freshwater crocodiles can literally snap them out of the air, while others wait patiently for an easier mark.

But the crocodiles pose a minor threat. They kill far fewer little reds than the orchardists did in the past, when they systematically poisoned or hunted the fruit bats. While their numbers are far down from what they were before European colonization, the little reds of Australia seem to be flourishing. As long as the forests they need to survive aren't felled for timber or cleared for agriculture and more cities, these winged wonders will continue to lift into the night, following the scent and sight of flowers wherever they lead.

ARRIVING AT THEIR CAMPSITE at sunrise, the flying foxes cling by their hind limbs in clusters (left). Males and females roost together for most of the year. A flying fox lives up to its name with great skill (above).

RACE TO

To many animals the paths of migration are deeply incised some-where in their essence. Whether memory, instinct, or their own mys-terious compass guides them, they follow the same routes through the same terrain. Often their migrations are a race for the blessings of food, or a race against the curses of weather. • To the **ZEBRAS** of northern Botswana, whose harems move with the seasons, rain is life. Just as ice is life for the **PACIFIC WALRUSES** who annually float north on it between the Bering and the Chukchi Seas. For a band of

SURVIVE

PRONGHORN in northwestern Wyoming, snow is the critical issue. They follow the melt north in spring, then race south in fall, trying to stay ahead of the coming snows. • In the canopy of the **BORNEAN RAIN FOREST,** food creates a marathon for many species, as they race toward a giant fig in fruit. And in a different way, the intricate food chain of the ocean keeps the Earth's largest fish, the **WHALE SHARK,** and the tiny zooplankton of the **DEEP SCATTERING LAYER** locked in a mutual dance that makes each day a race for survival.

An improbable vision in the middle of the Kalahari Desert, the swollen Okavango River spreads out every year to cover 6,000 square miles—the largest inland delta in the world. Like a vision of the old untouched Africa, this oasis of wet attracts elephants, buffalo, antelopes, zebras, and hundreds of species of birds. When the rains begin in November or early December, the animals start to disperse. For reasons only they comprehend, one group of plains zebras undertakes a daunting migration—into the brutal salt pans of the Makgadikgadi.

MOVING TO THE RAINS

The pans of northern Botswana were once part of a large inland lake that covered the Okavango Delta and the rest of the Kalahari. Lake Makgadikgadi disappeared 10,000 years ago, but the minerals it left behind now crust the saline-whitened surface of the pans. Those minerals and the need for them may explain the zebras' determined seasonal migration to the Makgadikgadi, whose very name means "vast lifeless land." But in the rainy season, even this lifeless land teems with life.

Crossing the roughly 150-mile mosaic of savanna grassland and woodland that lies between the Okavango and the pans will take the zebras anywhere from 10 to 20 days. Foraging their way southeast, they stop to drink from occasional puddles and seasonal water holes. The grass is poor, but the plains zebras are impressive eaters, able to digest and extract vital nutrients from grasses that other ungulates couldn't deal with. Their digestion also requires that they eat continually throughout the day and even during night.

At the northern edge of Makgadikgadi they reach the only consistent water in this dry world, the life-nourishing Boteti River. In a good year of plentiful rain, it flows freely. In dry years there are always water holes along its bed that promise rare, much needed moisture. Whatever the season holds, at the Boteti the Okavango migrants will probably find wildebeests and others of their own kind—some of the Makgadikgadi's permanent population of 15,000 zebras will be gathered at Boteti as well.

Within these larger herds are the harems that define zebra society: a male, his mates—as many as six of them—and their mutual offspring. The mares follow a strict hierarchy, the first chosen in the harem remaining the males' favorite. This matriarch takes the dominant role, leading the phalanx, her offspring right behind her, as the harem moves on. The rest of the females follow after her in order of their arrival in the harem, carefully adhering to a system as rigid as any human hierarchy.

Keeping an eye on his brood, the stallion often brings up the rear. Not only does he have to guard against bickering among his mares and occasionally even step in to protect a new female in the harem, but he has to stave off any competing males. This is the breeding season,

WATER IS AVAILABLE only irregularly in much of the zebra's range. The animals travel far each year to find it at traditional locations such as this one in Botswana.

ZEBRAS ARE WELL ADJUSTED to their migration routes in East Africa, where their perpetual goal is to find water and grasses.

When the zebras arrive at lakes and rivers, they often join wildebeest
herds that have thronged the river and its banks for the same goal.

and alert bachelors hover around, hoping to steal mares away from an established harem. The fierce fighting that often ensues when this happens is a pugilistic feat of kicking and biting that can result in serious injury. To the victor go the spoils—mating privileges with the fought-over mare.

For a mare pregnant from the year before, this is the time to foal. The young calf she drops stays close, nursing on her rich milk for months to come. But within the first month of the birth, the calf also learns to forage on the nutrient-filled grasses of the Makgadikgadi.

The mare's barks and whinnies and smell and pattern of stripes are distinguishing features, well known to her calf. Ironically, the very stripe patterns that make zebras distinctive to each other make them less individuated to predators. When lions scan a herd of foraging zebras, looking for an individual to hone in on, what they see is a weltering confusion of stripes, a camouflage that makes the hunting harder but not impossible.

Lethargic in the dense heat of the day, the lions often wait for darkness before they begin to hunt in earnest. The zebras know this, and as night falls they become more wary. They have to eat, even at night, but they leave the open grassland when they can and forage in the protection of brushier areas or woodland. They move more erratically and faster, to throw their predators off. But it doesn't always work. The lions are skilled stalkers, often cooperating together in a hunt. Sooner or later, they'll make a kill.

THE BREEDING SEASON means serious competition among zebra stallions for territory and females (left). Meanwhile, the zebras must continuously be alert for predatory lions (above). Following pages: Wildebeests and zebras run from danger at the Maasai Mara National Reserve.

As the wet season moves into January and February, the zebras graze the brittle grasses of the Makgadikgadi, rich in trace elements and proteins lacking in the lusher looking wet-season forage, and they drink their fill from rain-sweetened water holes. Whenever and wherever rain darkens the horizon, they move toward it.

But rain is a fickle master in Botswana. In years of plenty the zebras flourish, and the Okavango herd may extend its stay in the pans for up to eight months. During drought years the herd's numbers dwindle, and the animals may leave Makgadikgadi after only a couple of months. In a normal year, they begin the long march back to the Okavango in late April, but regardless of the month, when the water in the salt-pans area dries up or becomes too saline to drink because of lack of rain, the zebras will leave, moving quickly.

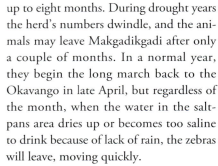

Their migration here at the beginning of the wet season had ensured them of full water holes along the way. They'd taken their time, arriving at Makgadikgadi after 10 to 20 days. They will make the return journey in a week or a week and an half.

The dry season and the equatorial sun have parched the savanna, and water is now a rare commodity in the Kalahari. But the zebras know where to find it, and they move fast toward it. The flooded delta is waiting for them, mile upon mile of sweet Okavango water.

A ZEBRA CALF stays close to its mother for months, recognizing her by voice, smell, and pattern of stripes (right). Crisscrossing footprints mark a prime location for nourishing grasses in Kenya (above). Following pages: A migrating zebra herd raises clouds of dust at twilight in Botswana.

To the human eye, the ocean seems to sweep across the horizon, its distances marked in surface miles. But to the life below that surface, distance is as vertical as it is horizontal, an up-and-down journey layered with changing currents and temperatures, light and dark. And every day through the ocean layers a phenomenal migration takes place, as urgent and extreme as any on Earth. A migration that keeps the ocean ecosystem balanced—and may well be a balancing agent for the health of the entire planet.

A BOTTOMLESS SEA

Across the globe, as Earth dips away from the sun and night creeps across the ocean, millions, perhaps trillions of minuscule animals begin their ascent from the ocean depths to the surface, propelling themselves upward as much as 1,500 feet in what may be the largest migration on the planet. Transparent and fantastical, they range from the size of a thumbnail to about six inches long. Collectively, they are called the deep scattering layer (DSL), not for any poetic reason but because their gas-filled swim bladders and drops of fat can scatter sound waves emitted by sonar devices.

The World War II sailors whose sonar screens detected them thought at first that they were the bottom of the sea, yet a bottom that rose every evening. By 1948, scientists realized that this "false bottom" was in fact a shape-shifting biomass composed of a wealth of zooplankton representing thousands of species, many as yet to be identified but others the larvae of shrimp, crabs, bizarre deep-sea fish, eels, jellies, and other minuscule organisms. The creatures of the DSL are on their way to their own foraging pastures—the phytoplankton, or microscopic plants, that live at the sun-filled surface of the ocean.

Their nightly migration is, according to one biologist, equivalent to a person walking 25 miles each way for a meal.

The phytoplankton and the zooplankton of the DSL anchor the food chain. In Belize, the food chain includes several kinds of snapper, who thrive in the mangrove wetlands along the coast and every spring make their own migration. They stream out of the mangrove shelters and across the Belize Barrier Reef, passing through the gateway of Gladden Spit, where the reef suddenly drops off a thousand feet.

On nights when the moon is just right, in full or quarter phase, cubera snappers school up about 200 feet below the surface and begin the annual mating dance. The sounds of their courtship rituals carry through the ocean waters, alerting predators like the bottlenose dolphin, who cruise in for an easy harvest. The dolphins aren't the only ones waiting for this moment. Whale sharks, the largest fish in the sea, have come to these waters specifically for the snapper spawn.

Despite their enormity—more than 40 feet long and 20 tons— these sharks are filter feeders, literally vacuuming the ocean's waters. With their enormous mouths open, they suck in and strain out

THE WHALE SHARK, largest of all fish, feeds on some of the sea's smallest organisms, zooplankton, which it vacuums from the water in immense quantities.

The zooplankton of the deep scattering layer make a nightly migration to the
ocean's surface to feed on organisms even smaller than themselves.

plankton and zooplankton—everything in their paths—then force the water out through their gills. The whale sharks migrate in, perhaps from oceans away, to join the annual snapper feast along the Belize Barrier Reef. But unlike the dolphins, they aren't interested in the mature fish. They've come for the snapper eggs and milt.

A milky explosion clouds the water as the female snapper release their eggs and the males their milt. The whale sharks, mouths open, move in for the kill, sucking up the released spawn. The fertilized eggs that survive become larvae dispersed by the current. Should they make it to adulthood, they will migrate back here for the annual spawning.

When the snapper spawning is over, the whale sharks will move off in search of other food. Left alone, they could survive to be a hundred, but in some parts of the world fishermen are anxious to harvest the huge fish for their meat and fins. The International Union for Conservation of Nature now lists the world's largest fish as vulnerable.

As the sharks roam long stretches of ocean, the tiny organisms of the deep scattering layer continue their nightly migration up to the surface to feed, then return to the deep ocean with the sun. As they do so, they take with them the carbon captured by the phytoplankton they devour, and they excrete or exhale the carbon into the ocean depths. Could their mini-migration be a linchpin in the climate complex, vital to protecting the Earth?

SEA TURTLES come into the Caribbean to breed after long travels (above). They share the warmer habitat with whale sharks (opposite). Following pages: Snappers, silverside fish, and a sea star share space in the underwater mangrove forest.

When first light brushes the canopy of the Bornean rain forest, a male gibbon high in the tree canopy sings the sun up. As the tropical light grows stronger and penetrates the dense deciduous top of this canopy—the highest in the world—gibbon families begin their aerial acrobatics, infants clutching mothers as they swing with skilled abandon from branch to branch. They move in the direction of the hornbills' calls. The birds are official harbingers of a forest feast, and the race is on to reach the bounty—a giant strangler fig in full fruit. In this leafscape of deep shadow and sprinkled light, where branches are intersections and sound is as important as sight, the fig is a keystone species, critical to survival.

FOR THE EYES AND NOSE OF GOD ALONE

Ten thousand, even 40,000 green globules, may pack the smaller branches of the fruiting ficus, some of the figs ripe, some still ripening. Birds, primates, and other animals all join in the eating orgy. Along with the gibbons, red leaf monkeys, long-tailed macaques, and the "forest people," the *orang hutans,* as the Malay call them, come to feast. The giant fig will fruit only once every two years, and when the great biannual burst occurs, the forest is waiting. The race is on to reach the fruit at its prime. The time to eat is now.

In these forests of dipterocarp trees, each plant and animal has mastered its own unique survival skills. A lot of the inhabitants of this cloud-scraping world never touch the ground, instead roaming a tree canopy that reaches 200, 250 feet into the sky. Bouquets of sweet-smelling flowers color the canopy top, yet finding food here isn't that easy, so finding it quickly when it becomes available is imperative.

The fig reigns supreme, an abundant giver and an opportunistic taker. "The topmost leaf-mass in the forests is largely composed of the foliage of trees of the genus Ficus," wrote naturalist Alfred Russel Wallace. A contemporary of Charles Darwin, Wallace too conceived a theory of evolution and natural selection. He was deeply impressed by the ficus on a visit to Borneo in the mid-1800s. He called the strangler fig's tactics "an actual struggle for life in the vegetable kingdom, not less fatal to the vanquished than the struggles among animals. . . . The advantages of quicker access to light which is gained in one way by climbing plants, is here obtained by a forest-tree."

Wallace was referring to the fact that, thanks to the animals that feed on its fruit and drop its seeds, the fig can sprout in the crotch of other trees high above the ground. At first it grows slowly as an epiphyte, sucking moisture and nutrients from the sun, rain, and debris on the host tree. Then it begins to develop roots that wrap around the

MORNING LIGHT AND MIST in the Bornean rain forest signal the time for gibbons, monkeys, macaques, and orangutans to begin the day's imperative search for food.

A PROBOSCIS MONKEY makes a flying leap through the lush forest of Borneo, where lianas twine around trees.

An orangutan dines on the tropical riches of a small Baccaurea tree
in Borneo's Gunung Palung National Park.

host tree, inch toward the ground, then tunnel into the soil, further sapping the host of water and nutrients. As the fig's roots thicken, they snake around the dipterocarp host, choking it. At the same time the fig sends up vines that grow toward the canopy, eventually covering the host's own leaves with its own and blocking out the sunlight. Over time the stranglers squeeze the life out of their hosts.

In an ironic way, the fig's aggressive patterns ensure the survival of a number of animals, because the tree is such a prolific giver. When in fruit, a fig feeds all kinds of animals, who are generally willing to share the largesse among themselves. The gibbons are the exception, occasionally chasing off red leaf monkeys and other primate competitors and never tolerating others of their own kind.

To a gibbon couple, territory is everything, shared only with their two to four young. The family claims their forest acreage by sending out a chorus of cries that can be heard a mile away, a warning that wards other gibbons off. The female's "great call" is worthy of a diva, trilling up and around two octaves and operatically changing mood and tone as she claims the fig feast for her own family. The smallest of the great apes, gibbons can brachiate from branch to branch with ease, quickly reaching the figs ripe for the taking. Once they've exhausted the supply, they swing on, moving quickly to another prime feeding spot. They have the advantage of speed over their primate cousin in the forest canopy, the lumbering orangutans, and they make sure to exercise it, because the orangutans are the true kings of the canopy.

The largest frugivores in the world, the male orangutans stand four to five feet tall and weigh in at over 200 pounds. It takes a lot of

APTLY NAMED PROBOSCIS MONKEYS (left) prefer the mangrove forests near rivers for their foraging. A mother and her baby (above) rest on a high branch in the forest canopy.

fruit to keep these apes satiated, but their size also means they have no natural predators. They go where they like, loners who rarely interact with other animals. And they are always after food. Their lives revolve around the search for it. If fruit isn't available, which is about half the time, they'll eat bark, leaves, stems, flowers, honey, even eggs, insects, and mineral-rich soils to keep them going. But it is fruit that keeps them happy.

Remarkably, orangutans seem to memorize an internal guide that leads them back to trees just at the time they are in fruit. But even without their memory, the sound of so many animals gathered together, calling, chattering, feeding, would alert them to the feast. The problem is getting there quickly enough.

Even though they're the world's largest tree dwellers they're careful climbers. Unlike the gibbons, they have no ability to swing gracefully through the canopy, or even to jump between trees and branches. The best they can do is use their weight to sway a branch back and forth, getting them close enough to the next foothold for them to reach it with their overlong arms. Sometimes they zigzag through the forest in their search for food, carefully weighing their options on the energy it will take to reach fruit against the rewards of how much caloric bang it will give them.

By the time they arrive at a fruiting giant fig, the feast is already on, with gibbons plucking and gorging on the ripe fruit, and hornbills expertly picking figs from a cluster, tossing them in the air, then spearing it on their bills. The orangutan settles for the sourer, unripe fruit, and a lot of what the big ape picks gets dropped. Landing on the ground, it is soon snapped up by the bearded pigs or deer or porcupines that have gathered below the fruiting fig.

As the sun sinks, the orangutan begins bending small branches toward each other, weaving them together then covering them with other twigs and branches to make a nesting platform for the night. To save energy and give itself a feeding advantage, the orangutan may build the nest right in the fruiting fig, so it wakes to easy picking.

Males sleep alone, but females sleep with their offspring until the young orangutan is six or seven years old. Though the young may nurse until they're five or six, they still crave fruit, just as the adults do. For the first year, the mother picks fruit to feed her baby, but it quickly learns how to grab the green bounty on its own. What it learns, it learns from its mother as the two make their lone way through the forest. Should two orangutan families encounter each other at the fig feast, the young may play together, but the females ignore each other.

The only real interaction a female has with another adult is when she's in estrus, about every seven to eight years. Then she looks for a male with the prominent cheek flanges and throat pouch that tell her he is mature and strong. If she's lucky, she attracts her preferred mate before a younger, less mature male finds her. The males make sure they stay out of each other's way, howling their "long calls" into the forest.

Far more sociable than their ape cousins—the gibbons and orangutans—are the monkeys of the rain forest. Easily the most striking among them are the well-named proboscis monkeys. Perhaps one of

the strangest looking primates in the world, these large, pot-bellied monkeys are found only in the Bornean lowlands, where they move together through the forests in raucous troops of up to 30 animals. Family groups and bachelor troops often congregate together in the evenings, sleeping near the rivers that weave through the rain forests. During the day, they move up and down through mangrove trees as they forage on leaves, covering long distances but never venturing far from the rivers.

In addition to their memorable noses, these monkeys have the distinction of being the only primate in the world equipped with partially webbed feet. They use their duck-like webbing to good effect when they need to dog-paddle across rivers, swimming almost silently, so as not to alert crocodiles to their presence.

Their cousins, the red leaf monkeys, or langurs, are creatures of the forest canopy, but they, too, are sociable animals, living in small troops of maybe a dozen, controlled by a dominant male. They chat-

ter together as they move agilely through the canopy, settling into the branches of a fruiting fig to enjoy themselves.

After a few weeks, the exhaustive supply of figs begins to give out, and the feasters move on, hoping for another bonanza. Instinctively, they know the next meal isn't guaranteed and the need for speed is real.

For all its lushness, the Bornean rain forest is an uncertain place. Weather here is subject to the roughly four-year cycle of the El Niño/Southern Oscillation, which can cause drought and attendant fires. The drought can be devastating to some species, but it encourages the dipterocarps to mast collectively. The forest canopy swells with flower, then fruit, creating a paradise that the Christianized Dayak people here say was meant "for the eyes and nose of God." The mass release of seeds literally carpets the forest floor and provides a feast for local fauna.

For centuries the dipterocarp forests that cover Borneo, the third largest island in the world, periodically nourished the entire ecosystem with their masting phenomena. In between mastings, the figs, also reliant on dipterocarp hosts for their very existence, fed the forest. But now that cycle is seriously threatened. During the 1980s and 1990s unbridled logging took a heavy toll. Today there are more restraints on logging, but legal and illegal deforestation continues, even in and around national parks. In some areas Borneo's lowland forests have been reduced by more than half, and the logging leaves behind soil conditions that don't encourage tree

THE BORNEAN RAIN FOREST is home to a great variety of agile animals such as the proboscis monkey (opposite, top and bottom). Their cousins, small red leaf monkeys (above right) and white-bearded gibbons (top left), are also inhabitants of the canopy of these lush forests. Following pages: A female Bornean orangutan carries her one-year-old offspring to safety.

regrowth. A change in the forest canopy seems to have an effect on rainfall patterns as well, further complicating things.

The rare plants and animals reliant on the dipterocarp forests are increasingly threatened or endangered. Territories for the big animals, like orangutans, are disrupted and discontinuous. And that has a feedback effect. The big apes themselves are seed dispersers, and as their territories are enclosed and limited by swaths of cultivation or logged

areas or forest fires, they disperse fewer seeds, changing the forest composition in subtle ways as yet unknown.

These Bornean forests are the only natural home of this strange and wonderful ape, and loss of habitat could spell its doom. The International Union for Conservation of Nature estimates that the orangutan population in central Borneo has been reduced by half in the past 60 years. Orangutans are listed as an endangered species. But loss of habitat is not the only threat. Young orangutans are prize trophies, taken for the illegal pet trade, and adult orangutans are sometimes hunted as well.

For now, their haunting faces continue to stare into the rain forest as if contemplating a world that is fading before their eyes.

DISASTROUS DEFORESTATION scars the Bornean landscape (above) and could doom one of the Earth's most wondrous apes, the orangutan (right), and other severely declining forest dwellers.

They are a dwindling, rogue band of pronghorn, survivors in a harsh land. Only an estimated 200 of them are left to make the biannual trek across northwestern Wyoming—the longest land migration made by any New World animal outside the Arctic. These uniquely North American animals—not true antelopes at all, despite sometimes being called that—once stippled the High Plains, their tawny goatlike bodies melding into wind-scoured hills and passes. Now they float across the landscape like phantoms from a grander past.

HOME ON A VANISHING RANGE

The pronghorn were here long before humans, millions of years ago, when saber-toothed tigers and cheetahs instead of snowmobiles roamed the land. With predators like the cheetah, survival went to the swiftest, and the pronghorn evolved to be one of the fastest land animals alive, with a stamina that keeps it going long after its predators have given out.

In recent history, the pronghorn's arch predator has been man. In 1881 alone, some 55,000 antelope hides were shipped down the Yellowstone River. Hunting laws enacted in subsequent years protected the pronghorn from such excess, and now about half a million roam Wyoming, their numbers about what the state's human population is.

But this one small band of 200 animals has settled into a precarious corridor of Wyoming, moving between the sagebrush basins of the Green River drainage south of Pinedale and Grand Teton National Park to the north. Because they live in one of the fastest developing and most sought-after patches of the American West, the group has learned to be intrepid, not about human hunters but about human obstacles. In their annual migration they have to maneuver through an obstacle course as demanding as any on Earth.

Winters are brutal on the High Plains, and the pronghorn depend on the patchy sagebrush that pokes above the snow for survival. Despite their thin coats, they seem adapted to the cold, and they can paw forage free of snow, if need be. The small band of 200 that collects on the Pinedale Mesa, in the basins of the upper Green River, are not alone. They share their winter quarters with other antelope, elk, and mule deer. By the time the raw High Plains spring begins to melt the snow, and new shoots of grass break through, the pronghorn are anxious to move on, following the spring green-up north through the Green River Valley.

Though much of this area is federally protected in national forests and Bureau of Land Management acreage—leased in the past for grazing—things have changed drastically here in the past ten years. Pinedale and the surrounding area have gone from a quiet Western world lightly trod upon by outdoor enthusiasts and lovers of Wyoming's rugged traditions to a mecca for companies that exploit

THE PRONGHORN, although an icon of the West's High Plains, is sparsely represented in this small Wyoming band.

These agile antelope-like animals are ideally suited to overland travel in dry climates,
but they are also adept at fording waterways.

the underground riches. Today a welter of natural gas rigs poke the skyline, as companies use a new hydraulic fracturing technique called fracking to extract the gas.

With the mining boom has come booming development, and more and more ranchettes and subdivisions interrupt the prairie and the migratory routes of animals. Joel Berger, a biologist with the Wildlife Conservation Society, estimates that in this corner of Wyoming and adjacent Montana, 78 percent of the pronghorn migratory routes have been lost, mostly owing to the fragmenting of the landscape.

The pronghorn evolved in a land free of impediments, and they've internalized a biannual path across the prairie, literally following in the footsteps of their ancestors, wherever they lead. They understand places where they can "see far and run fast," explains wildlife biologist Kim Berger. Running fast requires wide-open prairie, and in this corner of Wyoming, that is becoming as threatened as this small band of antelope.

The fragmentation goes on with every passing year, but new development is only part of the problem. For the pronghorn, the greatest peril probably comes from Highway 191, the main road through this part of the state. Crossing it can be as dangerous to the pronghorn as the crocodile-filled Mara River is to the wildebeests of the Serengeti.

They face the greatest peril at an ancient chokepoint where the margins of the Green and New Fork Rivers narrow their migratory path to just a quarter of a mile wide. Known as Trapper's Point, this bottleneck has spelled doom to passing pronghorn for millennia.

THREATS TO THE PRONGHORN abound in its High Plains home. Traffic crosses its traditional migratory paths (above), and drilling for natural gas (right) encroaches on its habitat.

Six-thousand-year-old fossilized bones of butchered antelope, including fetal remains, have been recovered here, proof that Native American hunters laid in wait for the pronghorn to thread through the narrowing.

Today cars, not arrows, make the kill as small bands of migrating pronghorn, from a few to a dozen, follow the melting snowline north in spring. The females are pregnant from the fall rut the season before. That just adds one more burden to the challenge of migration in a landscape that is increasingly more piecemeal, the prairie plowed under or developed or cut by barbed wire.

For the pronghorn, the wire fences can be a death trap, snaring them in an inescapable tangle. Since they're not equipped to jump obstacles, their only choice, except in rare cases, is to go under the barbed fences. But even a new "pronghorn safe" fence doesn't guarantee them safe passage.

They keep moving, following the Gros Ventre River and its small creeks northwest. If the pronghorn are lucky with cars and fences and weather and creek crossings, by late spring they've made it to the safety of Grand Teton National Park. No fences interrupt the high meadows and deep forests there. Most fawns foal in the relative peace of the park, finding a swath of short grass where they give birth, generally to twins. The newborns have a remarkable survival feature—they're virtually odorless, a feature that helps protect them from coyotes and other predators. And they have a second survival advantage: Their mothers fiercely protect them, running after coyotes that come too close.

Despite their mother's protection, the fawn mortality rate is abysmal, with only 10 percent surviving in some parts of the park. Coyotes, as wily as their reputation, manage to take down many of the young pronghorn, but wolves, bobcats, and even golden eagles will prey on the fawns. The young pronghorn have to be wily themselves, and they are up and foraging beside their mother within weeks.

The high-country season in the park is short. Winter comes early and unpredictably in northwestern Wyoming. By late summer, snow can dust the mountains. By early autumn, the pronghorn begin leaving the park in small groups, climbing up and over mountain passes as they make their way back to their wintering mesa below Pinedale.

In the spring they had followed the retreating snowline north but now they're in a race against the coming snows as they retrace the path south. If deep snow catches them in the higher elevations, they

RANGELAND FENCES are an omnipresent barrier to the pronghorn, which is not designed for leaping high. When it tries to squeeze under (opposite), it can be ensnared in a barbed wire death trap (above).

could be lost. That happened in 1993 to a group of late-migrating pronghorn. Trapped by an early snow, they never made it out of the mountains.

In a dwindling band of only 200 animals, just a few deaths matter. The ones who make it back will participate in the age-old ritual of the fall rut, the males staking out their territories and defending them, the females choosing the males with the genetic makeup that may help this small band survive.

If all goes well and the pronghorn outlast the relentless Wyoming winter winds and temperatures, they'll take on their fraught migration again next year. And if some environmentalists have their way, the small, resolute band of survivors will have an easier path to follow.

A National Migration Corridor protecting the pronghorn route has been proposed by the Wildlife Conservation Society and other environmentalists. Spearheaded by Joel Berger, the protected corridor would be only a mile wide and half a mile long and mostly on federal lands. Nonetheless it is being resisted by powerful groups. Without it, the Grand Teton pronghorn are almost surely doomed, a relic of a West where the buffalo roamed and the wide-open prairie promised to go on forever.

AWAY FROM HUMAN OBSTRUCTIONS, pronghorn can migrate on routes they have always taken (above), although fast-rushing rivers (right) sometimes present a natural challenge. Following pages: Dusk silhouettes a pronghorn at the Heart Mountain National Antelope Refuge in Oregon.

To the walrus, ice is life. The ice along the continental shelf off Alaska and Russia is the ground beneath the Pacific walrus's feet. An oxygen-breathing marine mammal, it relies on the ice as a place to rest, to give birth, to nurse, and to migrate. And with global warming, the ice is disappearing. The annual migration is becoming a race against time and distance, depth and disaster.

ICE TRAVELERS

Every winter thousands of these massive animals congregate in the northern Bering Sea. Here, the Anadyr Current, which churns up nutrients in the waters, and dynamic sea ice usually ensure good feeding conditions for the walruses. They need areas of discontinuous ice, where there are open leads and where prevailing winds push the ice away, creating polynyas—ice-free "lakes" in the frozen ocean. They find what they need southwest of St. Lawrence Island, and they come here to breed and wait out the Arctic winter, feeding on the clams that pebble the bottom of this shallow sea.

Diving 250, 300 feet, they scull through the soft sediment with their hind flippers, feeling along the bottom with their sensitive vibrissae, or whiskers, for sea cucumbers, crabs, worms, and their favorite food, clams. To get to the soft-bodied meat of the mollusks, they literally suction it or blow it out of its shell with jets of water. In a day a two-ton male can eat more than a hundred pounds, and a one-ton pregnant or nursing female consumes even more. The shells they leave in their wake and their furrowing along the sea bottom impact the ocean floor ecosystem in ways scientists are just beginning to study.

As the weak Arctic days grow a little longer and the pack ice around St. Lawrence Island begins to thin, the walruses ready themselves for the long trip into the Chukchi Sea. The bulls travel in herds, swimming to land and moving north. The females generally wait till April to make their start. Then, they crowd onto ice floes with other females and their young, forming nursery herds. The young calves were born the season before, but on the float north, pregnant cows will give birth to another generation. The pregnancy has been long, 15 months, and expectant mothers know that they'll need to be vigilant and wily to keep their newborn alive. For two years, the new calf will stay close to its mother, nursing and building up the protective layer of blubber that will help it survive.

The walruses' migration is a magnificent sight, their huge pinkish brown bodies weighing down ice floes as they drift north through the narrows of the Bering Strait, crossing the Arctic Circle into the Chukchi Sea. The ice has to be dense enough to allow these massive pinnipeds to get on and off as they dive for food along the way. When they clamber up onto the floe, they hook their ivory tusks—really overlong upper canine teeth that can reach three feet long—onto the ice then haul themselves up with their flippers. The cold water restricts their blood, making their skin an almost pale white not unlike the ice. But once out of the water and warming in the sun, the blood flow pinkens the skin again.

A PACIFIC WALRUS is exquisitely fit for its environment, with whiskers that sense food on the seafloor and tusks that are suited for battling and for hauling it onto ice floes.

WALRUSES CLAMBER UP a rocky shore to a seaside area
where they can rest before a tumultuous breeding season.

As more and more walruses arrive, the hulking mammals live in close quarters on a rocky beach before the males begin to seek mates.

To mothers of young calves, every dive is a challenge. For the first few days of the newborn's life, the mother is in constant contact with it, carrying the infant on her back into the water as she dives. Predators—particularly killer whale pods or, if land is close by, polar bears—may take advantage of any absence, grabbing calves that are too close to the ice edge and even tipping or cracking small floes to dislodge a calf. And predators are not the cows' only worry. Once the

calf is older and a bit more independent, it plays with other calves in the ocean and begins to learn the skills it needs to survive. But it can get separated from its mothers, floating away as the mother dives.

The bulls have none of these fears to contend with. Even the polar bear, apex predator of the north, won't take on an aggregation of bull walruses. So once they reach the Chukchi Sea, the males leave the ice behind and haul out on island beaches or along the coast. Thousands of bellowing bodies lie clustered together, draped over one another, individuals struggling up from the pack occasionally to dive for food.

During these summer haul-outs, they molt, losing their old coat and growing a new one. The males may doze and fast during the annual molting, but the females have no such luxury. Since they're nursing, they must continue to dive and eat and float with the ice even as they go through their own summer molt.

Land is a much more dangerous place for the calves than the ice floes, so the nursery herds prefer not to haul out. But in recent years the floating world so critical to the walrus young is dissolving. The summer sea ice in the Arctic has dramatically declined in the past few decades, especially in the shallow waters of the continental shelf. For

six of the past nine years, the shelf in the Chukchi Sea has been ice-free, sometimes for a week, sometimes for as much as two and a half months. This is a radical departure from typical summer conditions.

In the 1980s and most of the 1990s there was always some ice covering the shallow waters nearer the coast. Without ice, the walrus nursery herds face harsh choices. If they stay with the ice, they'll find themselves over the Chukchi's basin, where the ocean is too deep for them to dive for food. They'll be stranded on the ice and unable to feed themselves or their young. Though their bodies are streamlined and designed for aquatic life, walruses are not fast swimmers or able to cover long distances. Their time in the water is limited. Sooner or later, they have to haul out. Given those limitations, many nursery herds have been forced to choose the only other viable option—hauling out on coastal and island beaches already crowded with their male counterparts.

Taking a small, immature walrus calf into a crowd of bulls, each of which weighs 3,000 to 4,000 pounds, could be likened to taking an unleashed chihuahua to a mass rally. The mothers and their calves have

AS AWKWARD AS THEY ARE on land, walruses are nevertheless expert swimmers (above left and right) and are comfortably at home when foraging under openings in the ice (opposite).

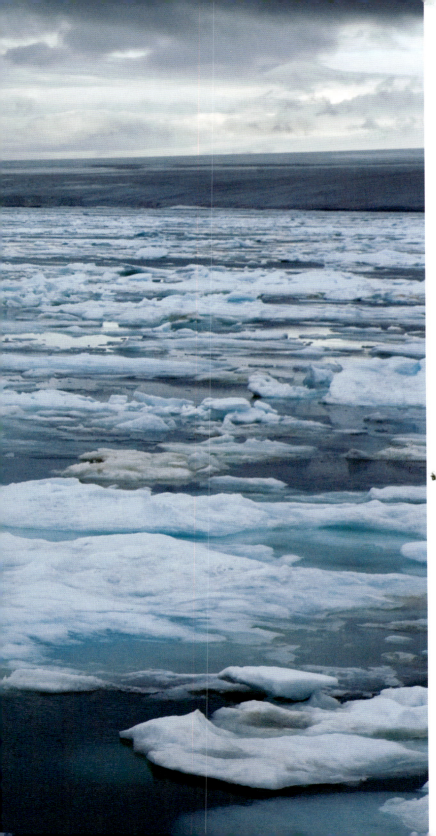

to clamber over a heaving sea of flesh. A walrus's body isn't built for climbing, and this kind of exertion takes a caloric toll. There's also the threat of the stray tusks of the males, inadvertently or intentionally causing harm. Before finding a relatively safe spot amid the horde of bodies, newcomers may travel half a mile without ever touching the ground.

Walrus reproductive rates are the lowest among the pinnipeds, with females giving birth about every two years. In the past, thanks to the nurturing care of the mother, most young made it to maturity and could live a good 40, even 60, years. But the changes caused by global warming are seriously threatening the survival of the young.

The demise of the walrus, tragic in its own right, could have profound effects on the Arctic world. The role these marine mammals and their feeding play in the health of the ocean bottom is probably critical. And the walrus is key to the traditional lifestyle of the Inuit people of Alaska. For centuries, they have relied on the walrus for food and for the skin that covers their boats. Even the animals' intestines are put to good use as rain gear. Each winter, when the walruses return to Alaska, the Inuit hunters are waiting for them. Now, fewer are returning, and the decline in the walrus population is be one more blow to this already struggling culture.

Exactly how many Pacific walruses are left on the planet is a matter of speculation. Surveys are now under way to determine their numbers, but as the ice goes, so goes the walrus. Without it, the annual spectacle of thousands of animals floating north through the Bering Straits will be no more.

SEA ICE IS CRITICAL in walrus habitat (left), and it is decreasing in the warming Arctic. A calf separated from its mother may make an easy meal for a polar bear (above). Following pages: A walrus calf rides its mother's back during a deep dive, and later rests with her on an ice floe.

FEAST OR

For migrants, their journeys are timed to follow food. In times of no food, famine makes migration a desperate business.   •   In Mali the last remnants of the great herds of **ELEPHANTS** that once populated North Africa are struggling to find food and water as they trek through the arid Sahel. And in the oceans, the greatest predator—the **WHITE SHARK**—is also in peril, its migratory path taking it into harm's way as it searches for food. Meanwhile, in a small lake in Palau, a unique marine creature, the **GOLDEN JELLYFISH,**

FAMINE

has evolved a daily migration that seems to keep it fed and out of harm's way. • One of the greatest migratory spectacles on the planet takes place along the **UPPER MISSISSIPPI RIVER,** as millions of birds fill the sky. Some of them are survivor species that are recovering from near extinction. In a world of ceaseless threats, their numbers are a celebration of endurance and hope—hope that future generations of migrants will continue to arc through the skies, roam the seas, and sweep across the land.

They glide like torpedoes of fear out of the depths of the ocean and into the imagination. White sharks, or "great whites," long believed to be the ultimate predator, are—like so much else in the marine world—still objects of mystery. Are they truly indiscriminate killers, these massive streamlined fish? How do they reproduce, and where are their breeding grounds? And where exactly do they roam? Most of these questions remain unanswered, but thanks to modern tracking technologies, the last question now has some answers.

OCEAN WANDERERS

One small spot in Mexican waters is a mecca for great whites and their prey. Lying west of the Baja California peninsula and at the intersection of the temperate and subtropical zones, the kelp-forested waters surrounding 22-mile-long Guadalupe Island are colored with reef life—tropical parrotfish and triggerfish, mola molas and butterfly fish, and a rare collection of marine mammals. Schools of dolphins and pilot whales leap off the island and the small Cuvier's beaked whale, almost never seen by humans, glides through the depths. Fur seals spin in the clear ocean, and after months at sea and a long migration south, elephant seals return to Guadalupe. For them, the volcanic island and its shores are more than a breeding and molting ground. In the past, this small hunk of rock, a dot rearing out of the ocean vastness, was the last refuge for the entire species.

Big, blubber-packed mammals, northern elephant seals had ranged through the northeastern Pacific before the 1800s. But early in that century they became an irresistible target for sealers. The oil rendered from the seals' blubber fueled lamps across continents, and the profits it brought led to wholesale slaughtering of the seals. By the

end of the 19th century, only an estimated 100 elephant seals were left on the planet.

This remnant colony hid among the rocks on Guadalupe Island's coast and managed to avoid most sealing ships. But even these secluded seals were occasionally exploited by hunters and museum collectors. Somehow, the tenacious group on Guadalupe survived against the odds. In 1922 the Mexican government conferred official protection on the elephant seal and declared Isla de Guadalupe a biological reserve—one of the first. Today, the steep, rockbound island and its current-swathed waters are protected as a biosphere reserve.

And the northern elephant seal has become an early and remarkable symbol of conservation success. More than 150,000 of them again range across the northern Pacific, about the same number that existed prior to hunting. But if governments have protected the seals from human hunters, they can do nothing to ensure them against natural predators, particularly the ocean's fiercest—the white shark.

Averaging more than 15 feet long and weighing a couple of tons, great whites roam the temperate and tropical oceans of the world.

AN ADVANCING WHITE SHARK typically means doom for any large sea mammal it approaches, even for huge sea elephants off Guadalupe Island off Mexico's Pacific coast.

WHITE SHARKS, known commonly as great whites, are masters of their ocean realm, swimming silently toward their prey.

Few potential victims manage to escape the predator's huge jaws.
Even mammals as large as elephant seals may lose chunks of flesh to a hungry great white.

Myths and misconceptions have fueled the public imagination, and the 1975 movie *Jaws* only added to the white shark's fearsome reputation. But after writing *Jaws,* author Peter Benchley disavowed his own creation, and before his death, he often championed them as remarkable animals. "I know now that the mythic monster I created was largely a fiction," he has said.

Among the many myths that have been dispelled about white sharks are their general movements. Surprisingly, they are often prowling the shallow waters where humans come to recreate, and yet white shark attacks are rare. Marine mammals are the sharks' preferred diet, and Guadalupe Island has plenty.

The sharks seem to know that as winter closes in, the elephant seals will head south, some finding their way to this spot 150 miles off the Mexican mainland. For the elephant seals the long migratory haul from the northern Pacific to Mexican waters has meant months at sea. They make continual deep dives as they move south, spending little time on the surface.

The bulls—with the great trunk-like proboscis that earns them the name elephant seal, or even sea elephant—arrive first. Like their larger southern cousins, many of whom clamber onto the breeding beaches of the Falkland Islands, the northern elephant seal bulls are intent on territory and mating rights with females. They tangle over position and dominance and wait for the females to arrive. Their massive bodies bear the scars of previous battles and worse. Some have hides seared with the marks of a white shark's teeth, and a few even

A NORTHERN ELEPHANT SEAL shows the bite scars left by a white shark that did not quite make this one a meal (above). Weary seals sleep head to tail in a pile after making landfall from a long migration (right).

have fresh wounds suffered in a recent attack by a great white. The wounded animals that have managed to make it ashore will probably heal. And the seals that have survived to a mature age have clearly learned the lessons of survival, whatever those lessons may be.

The white shark is a beautifully designed killing machine. Like all sharks and many other cartilaginous fish, it is equipped with a sixth sense, a result of organs called ampullae of Lorenzini. These small sacs, bundles of sensory cells connected to pores in the skin's surface, serve as electroreceptors, allowing the great white to detect the electromagnetic field emitted by another animal. So sensitive are these ampullae that the shark can detect half a billionth of a volt. This daunting detection device is only one way the white shark finds prey. It also has a keen sense of smell. Because smell travels better in the air than in water, the white shark will "spy hop"—raise its head out of the water—to pick up the scent of prey.

Then, if the prey is appealing, it moves in for the kill, using its powerful tail to propel it through the water. The prey that seems to appeal most in the waters off Guadalupe Island are the blubber-rich elephant seals. Efficient killer that it is, the shark doesn't waste energy that it doesn't need to expend. It moves in below and behind its prey, extends its fearsomely toothed lower jaw outward, and grabs its victim's hindquarters in one wrenching bite. Then, rather than wrestle with the seal as it dies of blood loss, the shark waits and lets the animal bleed out on its own. When the seal is dead, the shark again moves in for more, grabbing at the corpse and shaking it side to side as it tears off great hunks of flesh with its serrated teeth. Again, the shark's remarkably efficient system works to its advantage. Its metabolic rate is low, and it expends little energy in swimming. It can also ingest a large amount of blubber in one feeding and then live off that for as much as a month at a time.

The sharks' feeding opportunities increase as the mating season on Isla de Guadalupe progresses and the female seals start to arrive. They have been at sea even longer than the males when they reach Isla de Guadalupe. For some young females, this is their first season on the breeding grounds. Their inexperience makes them easy marks for the sharks, and for the eager elephant seal bulls.

As spring approaches, the mating season draws to a close, and the seals begin to disperse, moving north again into the waters of the North Pacific. The sharks, too, set off across the Pacific Basin. Eventually, many will make their way to one remote spot in the Pacific, about midway between Baja California and Hawaii.

ADULT AND YOUNG elephant seals haul out on a Guadalupe Island beach during the breeding season (opposite). In the waters just offshore, sharks are waiting to strike (above).

The researchers at the Monterey Bay Aquarium who discovered this mid-ocean meeting place in 2002 dubbed it White Shark Café. Using satellite tracking, they monitored shark movements and found that tagged males, females, and juveniles in the Pacific Basin all made the long journey to the café. Once there, the animals loitered, making dives down to a thousand feet. What is down there and why the animals congregate at White Shark Café is yet another mystery.

What is known is that the white shark is an apex predator and that it is vanishing from the oceans. Estimates put the decline in the global population during the last 50 years at somewhere between 60 and 90 percent, but exact figures of either the true population of white sharks or their actual decline are hard to determine.

The shark's slow growth rate and low fecundity make population recovery all the more difficult, and the International Union for Conservation of Nature lists the animal as vulnerable. Still, white shark fishing, illegal in many waters, goes on, and bycatch and loss of habitat only exacerbate their plight.

Far from threatening humans, these remarkable fish are now threatened themselves with annihilation. And in the ocean's interdependent environment, the loss of the apex predator will surely have dire consequences.

A WHITE SHARK slices through the sea surface attended by gulls seeking potential leftovers from the shark's prey (above). Flocks of adult and immature western gulls (right) search for small fish stirred up by the herd. Following pages: White sharks search for meals off Guadalupe Island, Mexico.

In north-central Mali, the Sahara's sea of sand laps insistently at the Sahel, an arid band that swelters under the African sun, its rainfall sporadic, its boundaries shifting with the dunes. Here, where daytime temperatures routinely reach 120°F, the last desert elephants have managed to survive by staying on the move. Making a 300-mile circuit every year—the longest known elephant migration—they have learned to outlast the weather, their long memories taking them from watering place to watering place. But as the climate becomes more fickle and human demands on land and water increase, these desert nomads face an ever more uncertain fate.

ON THE EDGE OF SURVIVAL

In earlier times, elephants wandered the whole of Africa, from the Cape of Good Hope at its southern tip to the shores of the Mediterranean. As recently as the late 19th century, pachyderms were still plentiful in West Africa. But in the century that followed, the territory that elephants—the Earth's largest surviving land animal—need to survive shrunk significantly. In the Sahel, it decreased to only a small percent of what it had been, as the human population there increased fivefold. Now 40 million people eke out their own living on this fragile "shore" (the meaning of Sahel in Arabic). And the remnant band of desert elephants, a scant group of only 350 to 450 animals, must live among them and their cattle and goats, sharing water and forage and open space in a corner of Africa that has less and less of all of these all the time.

The corner is actually the Gourma region, a shallow depression south of Tombouctou where scrub brush and sporadic forests make scant patches of green among the sand plains and dunes, and where wetlands and watercourses promise moisture even in the endless eight months of the dry season. At the beginning of the dry season, the elephants of Tombouctou, as they're sometimes called, are concentrated in the marshes at the north end of their range. The males, generally loners, move to their own tempo as they search for water and forage. Families of females and juveniles stay close together, rarely more than a few feet from other members of their group. They lumber single file along watercourses and thickets that shade them from the sun, following behind the matriarch who sets the pace for the group.

The older the matriarch, the better off her family. More experienced females know how to lead effectively and how to protect the newborns and young through their long, dependent childhoods. Within an hour of birth elephant calves are on their feet, but they stay close to their mother, nursing and taking haven under her massive body. For years, they'll look to their mothers and the other females in the family for their care. Even the juvenile females pitch in as

THE ELEPHANTS of Mali advance through the arid Sahel during their annual 300-mile circuit, which is closely linked to seasonal availability of water in the desert.

A LONG, DRY STREAM BED and the dusty, eroded remnants of its former tributaries
are symbols of the arid desert in Mali, where overgrazing has left much land unproductive.

Mali elephants must travel in a perpetual migration across the arid Sahel region
in search of sufficient food and water.

"aunties," helping with the young and teaching them how to live in the harsh Mali desert.

The blessings and dangers of mud are an early lesson worth learning. Wallowing in the chocolate-brown moisture is more than indulgence; it is an essential way to cool down, to protect against the piercing sunlight, and to remove annoying parasites. But mud can also be a slippery trap, capturing an inexperienced young elephant in a water hole. When that happens, the female clan coalesces around the young one, gently leading it out of the muck.

The females are generally quick to surround and help other members of the group who are in trouble. They're demonstrative, communicative animals that touch one another and call to each other frequently. They can hear well below the decibel range picked up by human ears, and their low infrasonic bellows can be picked up by other elephants several miles away. Their keen senses even detect the seismic waves passed through the ground by the sound of other elephant feet, alerting females to another family group or an approaching male.

A female in estrus lets her condition be known by giving out a strong call that bulls can hear a couple of miles away. The mates most preferred by females are mature bulls who are in a state of musth. A bull in musth is in a sexually aroused period himself, and the mating is almost always successful. Once pregnant, the female will carry the developing young for 22 months. In all, it will be four years before she is sexually receptive again. That makes every calf born critical to the survival of the family—and in the case of the Mali elephants, to the entire band.

As the dry season progresses, water and forage become increasingly hard to find, it is the infants that suffer most when they have to travel long distances searching for sustenance. Now, the elephant families as well as the lone males begin to head west, moving quickly toward the one spot that even in the worst of the dry promises water—Lake Banzena.

Lake Banzena has long been an oasis for all life in this part of the Sahel. Like the elephants, the Tuareg and Fulani people here frequent Banzena, feeding and watering their livestock at the lake's edge. In recent time overgrazing has bared the earth around the lake, and wind-blown sand and dust silt its waters. Fewer trees and shrubs interrupt the plains beyond it, leaving the elephants less forage and fewer places to take shelter from the daytime sun. Stands of dead trees, "woodland cemeteries," are testament to changing patterns in climate and the human footprint on the land.

MATRIARCHS LEAD FAMILIES of related females and their young in the endless search for forage (opposite). A group bathes and drinks during a refreshing pause (above left). The adult females keep careful watch on young in case they become stuck in the mud (top right). Following pages: A giant dust storm sweeps across the Sahel.

For centuries the Tuareg and Fulani crisscrossed the Gourma in their own annual migrations, living easily with the elephants. Nomadic herders who depended on their animals for survival, they moved with the rains, just as the elephants did. They even followed the great animals, knowing the matriarchs would lead their families—and them—to water. But since the 1970s these herdsmen have become more and more settled, and that pattern has shifted the age-old balance in the Sahel. Their villages are near water points, and their willingness to share water with the elephants depends on just how much the rains have given.

Typically, clouds begin to gather in May or June. But nothing is typical in the Sahel these days, as climate change and deforestation have disrupted predictable patterns. Dust storms blow up like tornadoes of sand, often in front of the coming rains, and drought is a constant threat. In 2002 the Gourma had the lowest rainfall in 50 years, yet the next year saw the highest precipitation since 1965. Then in 2009 drought again baked the land, the worst in a quarter century. Even Lake Banzena dried up, leaving only a few mud pools filled with dead cattle and struggling fish. What little water was available from two pumps set up by the government had to be shared by the local herdsmen's cattle and the Mali elephants. The cattle came first. Several elephants died during the drought, but somehow most of the dwindling band survived.

In any given year, when the rains finally break open the bleached earth and give them the freedom to travel, the elephants make their move immediately, all of them heading south toward the wet-season

A BABY ELEPHANT stays close to its mother, never concerned about being stepped on (above). As a herd trudges onward, two of its members are more interested in each other than in continuing the trek (right).

feeding grounds. The route the elephants follow once took them through open ground but now settlements, some 200 new ones, crowd their range and block their direct path to the water in the south. The settlements create choke points—narrow corridors the elephants have to negotiate to make their migration. They must pass through "La Porte des Éléphants" ("The Elephants' Door"). This is one of the only breaks in the Gandamia Escarpment that is not blocked by human settlement. It is the vital opening that will take them to the lush grasslands of Boni in the south. How quickly they reach these green pastures is critical to their survival.

And can they survive? Predictions for the future of the Mali elephants are grim. The nonprofit Save the Elephants, founded by pioneering elephant scientist Iain Douglas-Hamilton, warns that these northern elephants could be gone in a dozen years, given the pressures on them from climate change, habitat degradation by livestock, and other human activity. Douglas-Hamilton has worked to provide them with emergency water in the recent drought, but he and other conservationists know that the elephants' fate lies with the Malians, who have long revered their elephants. Popular songs have even been written about them, and Tuareg, Fulani, and other indigenous peoples in the Gourma have always shared their land with these enormous partners. With more humans and fewer resources to go around, the partnership may need to shift more in the animals' favor if the elephants of Tombouctou are to be saved.

AN EXTENDED FAMILY of elephants arrives at the wetland oasis after a long journey through the dry savanna (above). Clouds and rain typically come in May or June, breaking open the scorched earth and giving the elephants vital wet-season vegetation (left).

The mothers and other adults in a herd pay close attention to the very young elephants,
which sometimes need help to climb up from sticky mud.

A primordial dreamscape of floating ephemera, Jellyfish Lake is home to the stunning, brainless little beauties that gave it its name—golden jellyfish. These small creatures are their own miracles of engineering, living symbiotically with the algae they carry in their bodies. Being careful caretakers, they follow the sun in a daily migration that feeds their passengers and ensures their own survival.

FOLLOWING THE SUN

Their lake home, a petri dish of sorts, is relatively recent in geologic terms. It formed only some 12,000 to 15,000 years ago, as the last glaciers retreated, sea levels rose, and the limestone depressions on several Palauan islands filled with seawater. These marine lakes became their own isolated worlds, invaded only by the tidal flow from nearby lagoons that leaked in through tunnels in the limestone.

"Each of the salt lakes presents an unique natural experiment in the organization of food webs—the interdependence of life forms," according to the Coral Reef Research Foundation. In the lake on the island Eil Malk, a strange and marvelous kind of creature appeared—the golden jellyfish, *Mastigias papua etpisoni*—unlike any other lifeform on Earth.

As with other members of the phylum Cnidaria, the golden jellies are simple creatures with an elementary nervous system, or "nerve net," instead of a brain. As they evolved in the isolated world of their particular marine lake, they adapted to what the lake and the tropical setting had to offer—sunlight, lots of rain, shelter from the wind, little change in temperature, and most important the lake's algae-like zooxanthellae.

The "zoox" are remarkable single-celled creatures that live symbiotically inside their hosts. In other marine environments, zoox feed corals, sponges, and giant clams, but here their partnership is with the golden jellies. The zooxanthellae are sun eaters, photosynthesizers, whose excretions feed the jellies three-quarters of what they need to survive; the other quarter they get by ingesting zooplankton in the lake water. In turn, the jellies must be porters for the zooxanthellae, taking them daily up, down, and across the 1,500-foot-long lake.

Every day before dawn the golden jellies—as many as ten million of them, depending on the year—begin their ascent toward the surface. As they make their daily migration, their bellies contract, pushing water away and thrusting them forward by jet propulsion. When the medusae—the jellies—reach the surface, they densely pack the western end of Ongeim'l Tketau, the Palauan name of the lake. Then around 6 a.m., when the predictable tropical sun rises, its shafts hitting the eastern end of the lake, the medusae swim diligently in that direction, filling the other end of the lake with their golden bloom.

Once there, the medusae rotate their bodies so the algae they carry can bask in the morning glow. Otherwise their swimming is no more than milling about, avoiding the tree-shadowed edges of the lake, not just because of their lack of light but because of the predators that line the edge.

GOLDEN JELLYFISH of Palau receive their namesake color from algae-like, single-celled organisms named zooxanthellae, which live within a jellyfish and provide it with the energy required for life.

GOLDEN JELLYFISH of Palau migrate daily to a lake's sunlit surface, where algae-like organisms within their bodies transform the light into energy for the jellyfish.

At nightfall, the jellyfish make a return trip far down into the lake,
where they can absorb essential chemicals found only in deep layers of the water.

Anemones, relatives of golden jellies, wave their innocuous-looking tentacles in the current, hoping to catch a wayward jelly as a meal. These white, medusa-eating anemones are the greatest threat the jellies face. Some fall victim, but most of the millions make it safely through the early afternoon, waiting until the sun passes overhead and into the western sky. Again, they're on the move across the lake to congregate where they began the day, feeding their algae parasites light for as long as possible.

They may also be "biomixing," say scientists Michael Dawson and John Dabiri—that is, creating enough turbulence to churn nutrients in the lake waters, bringing the zooplankton they ingest to within reach of their stinging tentacles. "Biomixing may be a form of 'ecosystem engineering' by jellyfish," says Dawson. He and Dabiri are even looking at how the turbulence jellyfish create throughout the world's rivers, lakes, and oceans may affect marine currents and the flow of vital nutrients and chemicals in the food chain.

As shadows fill the lake surface, the medusae no longer have light to orient their movements across the lake, so they begin their nightly vertical passage down through Ongeim'l Tketau.

Like other marine lakes in Palau, the golden jellies' home is a rare, stratified world. The upper layer of the lake is oxygenated, and the medusae live and thrive in this world. But they also descend about 45 feet in their nightly maneuvers. They get as far as a well-demarcated change in the lake chemistry, a chemoline. Below this point the waters

THE TROPICAL ISLAND nation of Palau in the northwestern Pacific Ocean is a geographic parade of lush islands and lagoons (left). Among its environmental gems is Jellyfish Lake (above), an extraordinary ecosystem that is home to millions of golden jellyfish.

are anoxic, or oxygen-deprived. And at the top of this layer floats a dense cloud of purple sulfur bacteria that live by photosynthesis. They lie like a blanket above the lower depths of the lake, absorbing any light that might reach it.

The darkness below the chemoline is a deadly chemical stew of hydrogen sulfide, ammonia, and phosphate. Human visitors to the lake, who come to snorkel, are warned to avoid the toxic depths. The medusae instinctively know to do that, stopping in the chemoline, where their algae partners feed on necessary nutrients.

In 1997 and 1998, the golden jellies' algae suddenly went through a massive die-off. Scientists speculate that it was caused by an El Niño weather event that raised the water temperate beyond tolerable levels for the zooxanthellae. Without the algae, the golden jellies perished, disappearing altogether from Ongeim'l Tketau. But after a year and a half, they were back. Medusae polyps, an early phase in the jellies' life cycle, had survived the change in temperature and were able to spawn after the lake cooled again. In 2005, the medusae had a record year, with over 30 million turning the water to gold as they pursued their daily migration up, down, and across Jellyfish Lake.

Are they in fact as important to the lake as the zoox are to them? "The real question," says Michael Dawson, "is whether the lake keeps the jellies alive . . . or the jellies sustain the lake."

SEA ANEMONES are a prime enemy of the golden jellyfish. Their tentacles are ever alert to capture and they will consume a jellyfish that wanders into their grasp (above and right). Following pages: The jellyfish make their nightly pilgrimage to feed the tiny organisms that live symbiotically inside their hosts.

"**O**ne swallow does not make a summer, but one skein of geese, cleaving the murk of a March thaw, is the spring," the great naturalist Aldo Leopold wrote. In his corner of the American Midwest, where the upper Mississippi River lies on the land like an aviary highway, vast caravans of birds make their way north and south in a seasonal migration that fills the air with honking, bugling, and wing beats.

WINDS OF CHANGE

About 40 percent of the migrating waterbirds and shorebirds on the continent travel through this corridor. Here, no mountains block their flight path, and the river itself runs like an open highway, funneling migrants to their distant destinations. From the high Arctic to the southern tip of South America, the birds come and go, their course sometimes veering away from the river, sometimes taking advantage of its plenitude—forested islands, shallow backwaters, high bluffs that provide good vantage for prey or nest sites for young. The Mississippi has etched these bluffs over eons, carving a bed into an ancient plateau known as the Driftless Area, because no glaciers or glacial drift ever altered the ragged terrain.

Until the early 20th century the river flowed freely through the Driftless Area, coursing over waterfalls and rapids as it raced toward the Gulf of Mexico. But in the 1930s, the Army Corps of Engineers began building dams and locks to smooth the upper Mississippi's way and make shipping more feasible. Surprisingly, those dams were a boon to the bald eagles. Fish stunned by the turbulent waters of the dam make easy prey in the cold months, when food can be hard to find. And the dams keep stretches of the river ice-free. By February, the local bald eagles are joined by migrant eagles following the thaw

north. They ride in on thermals, currents of warm air, and they feed on the abundant food along the river.

Official symbol of the United States, the eagle was once abundant across the country, but in the 20th century they were decimated by hunting and the pesticide DDT. The national bird actually had a bounty on its head in some states early in the century, because its predatory instincts interfered with human interests.

By the late 1930s, early conservationists fought against hunting the mighty hunter, arguing that the eagle's very survival was in question. In 1940 the Bald Eagle Protection Act was passed by Congress; in 1972 the U.S. banned DDT as an agricultural pesticide. Scientists had come to realize that DDT impacted both the fertility of adult eagles and the viability of the eggs that did get laid, making the shells too thin to withstand incubation. Now, almost 40 years later, the population of bald eagles has swelled to an estimated 300,000 birds.

"Eagles," wrote the poet John Keats, "may seem to sleep wing-wide upon the air." Their circling ascent into the sky is an impressive sight as their broad, outstretched wings allow them to soar on thermals with little expenditure of energy.

IMMENSE FLOCKS of avian migrants funnel through the Mississippi Flyway every day during the birds' twice-yearly travels between wintering and breeding grounds, offering a spectacular panorama of life.

They may move only to a nearby lake or river for the night,
but on a spring day they will depart on a long migration northward to their breeding grounds.

In spring the bald eagles migrating along the Mississippi corridor follow the river or its Missouri River tributary north to their breeding grounds in Canada. Generally, they begin their traveling in midmorning, when the thermals lift them effortlessly along. Without the thermals or the updrafts of air forced up from the high bluffs along the river, they're forced to flap-fly, a cumbersome, energy-expensive locomotion.

The eagles move quickly in the spring, anxious to stake an early claim in the competition for a mate and a nest site on the breeding grounds. The turbulent spring weather helps carry them along, as they move over a still-frozen landscape.

With the fish they prey on locked in ice, they may go for days at a time without eating. But as the grip of cold loosens, the ice crusting the river begins to give way. As it breaks up, the carcasses of dead shad and other fish bob to the surface— a seasonal exhumation that provides a feast for scavenging eagles on the upper Mississippi. If the fish the eagles prefer aren't available, as spring progresses there will be a river full of migrating waterfowl and other birds—and their chicks—to choose from.

Noisy flocks of mallards are among the earliest of these migrants to arrive in the Midwest. These highly flexible, ubiquitous waterfowl can adapt to all kinds of situations and climates, which has made them the most abundant dabbling duck on the continent. Mallards may stay put and never migrate, nesting in cultivated fields or even suburban backyards, but many take to the wing in late winter. Helped along by tailwinds, they follow the snowmelt and the availability of open water north to nest. Some flocks of mallards will keep going all the way to

BALD EAGLES MIGRATE along the Mississippi corridor in the spring, toward breeding grounds in the northern United States and Canada, finding plentiful food resources, and sometimes joined in feasting by common ravens (left and above).

Alaska before they nest, but others start their brood wherever they find suitable cover for a ground nest and plenty of food and water.

The rituals of courtship between mallard drake and hen normally have begun long before the migrating ducks arrive at their nesting grounds. A female may well have chosen her mate the fall before, impressed by a drake's balletic, ritualized swimming display. Or she may wait and select a mate during the migration north, perhaps anointing him with a touch from her bill.

Some hens even play a coquettish game to evaluate a potential mate: They may incite a drake male to follow them by displaying as they swim, then head toward their preferred mate as the other suitor follows. The hen's staccato quacking provokes her lover to attack her pursuer. The two drakes, their signature bright green head extended, square off in a performance of prowess.

The mallards are not the only birds with mating on their minds. As spring lengthens and warms the upper Mississippi, other arrivals begin to nest. Peregrine falcons, among the most talented hunters in the natural world, find the bluffs above the river a good place to begin a brood and the river environs a rich hunting ground.

The falcon's name itself, peregrine, means wanderer, and these raptors are found worldwide. Magnificent fliers, they are capable of making one of the longest distance migrations of any bird in the world, traveling 15,500 miles every year between summer breeding grounds on the Arctic tundra and nonbreeding sites in South America.

The peregrines are another success story, comeback birds who, like bald eagles, were decimated by DDT in the mid-20th century. Although the falcons did not ingest the poison directly, the chemical had infiltrated the food chain and seriously hampered peregrine reproduction. Most peregrine pairs did not begin to produce viable young again until the late 1970s, after the pesticide had been banned for several years.

By the century's end, the duck hawk, as the peregrine is sometimes known, had made a strong recovery, thanks in large part to scientists, falconers, and other admirers of the bird who began captive-breeding programs to perpetuate the species. Captive-bred birds released into the wild were "hacked"—provided food until they learned to hunt on their own. Some were released into urban settings, and peregrine populations in a number of cities are growing, the skyscrapers and

GLORIOUSLY COLORED wood ducks (above left) and mallards (above center and right) are among millions of waterfowl that migrate along the Mississippi Flyway. Huge mallard flocks (opposite) rush northward in spring to their breeding range. Following pages: A mallard pair swims in a cypress swamp near Lake Charles, Louisiana.

Like torpedoes of doom, peregrines dive out of the sky, locking on their targets. They have been clocked at speeds of over 200 miles an hour as they race toward a bird in flight and knock it with their clenched talons. The peregrines have a harder time following behind a bird in level flight. Then the prey has a chance of surviving. And the falcons rarely try to strike birds on the water, so the ducks congregating on the upper Mississippi are relatively safe as long as they're not on the wing. If a duck in flight is being pursued by a peregrine, it may drop to the water to protect itself from attack.

Whatever prey the falcon takes, it shares with its young—generally a clutch of three to five nestlings whose long, wailing cries keep the parents searching for food. The young ones fledge and begin to hunt on their own at two or three months old, but until then the parents are ferociously protective. Any potential threat to the chicks or eggs will be met with a vengeance. The peregrines will even chase off a bald eagle that strays into their territory.

high bridge girders of the urban landscape taking the place of the natural cliff faces the peregrines use in the wild for nesting and hunting vantages.

Peregrines are now found across the continent, some migrating and others full-time residents. The migrating birds have an impressive homing instinct, making their way back to falcon aeries that have been used by previous generations. Ornithologists have found that some of these sites—"scratches," as they're called—have been used continuously for hundreds of years by successive generations of falcons.

On the upper Mississippi mating pairs find good nesting sites in the ledges and holes in the river bluffs. Males, who are substantially smaller than females, arrive first and use their greatest talent to court a mate, making phenomenal diving displays in front of the nesting bluffs. Once mates are chosen, the female selects a nest site in the bluffs. From this high vantage the couple can also scan the activity below. These raptors are quick to take full predatory advantage of the seasonal parade of migrants coming and going on the river.

In early summer new wings appear over the dammed waters of the upper Mississippi as the mayfly hatch begins. Swarms of these ephemeral insects hang above the river in black clouds, to the dismay of people along the river and to the delight of songbirds, which stuff themselves during a short-lived feast. The mayfly nymphs have spent two years or more underwater, feeding on other aquatic insects as well as algae and plankton and being themselves food for fish. When the time is right, the nymphs swim to the surface or crawl onto nearby rocks or plants. Often within minutes they molt into a winged, adolescent stage, then molt again into adults. So vast are the swarms

A PEREGRINE FALCON is not only a mighty hunter but also a dashing figure, its head marked prominently as if it were wearing a dark helmet (opposite). Fluffy white falcon chicks (above left and right)—"cute," humans might call them—offer no hint that they will someday be streaking, power-diving predators.

of these newly airborne insects that they show up on national weather radar. But only briefly.

Once they emerge from the river, the mayflies have just one duty to perform—to mate in the time left to them. That time is short, only a day or two at most. (Their scientific order name, Ephemeroptera, means "short-lived wings.") Once their eggs are fertilized, the females lay them over water then die. In recent years, the summer hatch of mayflies on the Upper Mississippi has been explosive—a sign of the river's ecological health.

And there are other signs. Waterfowl whose numbers were once on the wane now fill the skies and make migratory stopovers in the Mississippi River Valley. The distinctive V's of Canada geese in flight arrow through the spring dusk, their honking a harbinger of changing seasons. The flocks are composed of assorted individual geese, family members, and mated pairs who will remain together for life. The big brown bodies of the honkers bobbing on the upper Mississippi may be juxtaposed with the arctic whiteness of snow geese or tundra swans bound themselves for tundra breeding grounds.

Like all the spring migrants, the geese are anxious to press on. Riding high-pressure systems north, they will briskly make their way toward breeding grounds in Canada and Alaska. And like other migrants, some of the Canada geese will find all they need here in the upper Mississippi River Valley and will stay, building a feather-lined nest for their clutch of five to seven eggs.

The newly hatched goslings are born to swim. They take to the water right after birth, often following in a line, with one parent in the lead and one bringing up the rear. Readying themselves for the migration ahead, they eat almost continuously, feeding on aquatic plants and on river-bottom nutrients. The geese may also take advantage of ripening fields, making them a pest to farmers. But by late summer, even resident populations will fly north, if only a short distance, to molt. Then it will be time for the flight south to winter grounds.

A strange site for the center of the continent, flocks of American white pelicans, too, coast in for ungainly landings on their way to the inland lakes and prairies where they will breed in boisterous colonies of several hundred to several thousand birds. As fall moves in, the big ponderous birds will be back again, on their return trip to the lower Mississippi and the Gulf of Mexico's bays and estuaries.

As the rivers and lakes freeze in fall, the ice and cold will chase all the migrant birds south. Without the urge to breed pushing them, they'll make a more leisurely flight to their wintering grounds, filling the skies and river again with the great migratory spectacle—from

MAYFLIES ARE A STUDY in striking contrasts. Their nymphs spend two years in the water, but the adults (top right) emerge, mate, and die within a day or two. All are a superabundant food resource for frogs (above left) and migrating songbirds at the upper Mississippi region's rivers and lakes (opposite).

the heavy forms of the pelicans and geese, to the plump blackish American coots that are easy prey for migrating bald eagles and peregrine falcons. And gliding majestically on the daytime thermals, huge "kettles" of sandhill cranes will sometimes float like fall clouds above the horizon, bugling as they make their biannual migration. Among the oldest birds on Earth, their fossils dating back more than nine million years, they are a testament to the longevity and endurance of the migrants.

In other ways, the avian spectacle on the upper Mississippi is a sign of changing times and greater stewardship. In the previous century, many of the birds that now traverse this ephemeral highway were in serious trouble, hunted and poisoned both advertently and inadvertently. The giant Canada goose, *Branta canadensis maxima,* now so numerous that it can be a serious nuisance to city dwellers and farmers alike, was nearly extinct by the early 1900s. By 1940, the sandhill crane, *Grus canadensis tabida,* had been extirpated from large areas of its breeding range, and in that same year, the Bald Eagle Protection Act was enacted, an early and enlightened attempt at conservation. Still, eagle and peregrine falcon populations, as well as other migrant birds, continued to decline precipitously.

By the 1960s, much of the continent's birdlife was disappearing, and a former government biologist warned of a "silent spring." Rachel Carson's 1962 book of the same name became a groundbreaking classic and an inspiration for the incipient environmental movement. In it she

WHITE PELICANS (above and right) are among the largest Mississippi Flyway migrants. At an average weight of 16 pounds, they rely on a nine-foot wingspan to enable flight at very high altitudes. They use a unique fishing technique, cooperatively herding fish into small areas on the surface where the pickings are easy.

described how the production of synthetic chemicals that had begun to reach large quantities in the 1940s had "become an appalling deluge . . . daily poured in the nation's waterways." These pollutants, she said, were also infiltrating groundwater. The last chapter of Carson's *Silent Spring* begins, "We stand now where two roads diverge."

For many animals we humans still must choose which of the two roads we will take in our activities if we want to preserve migratory corridors and allow other creatures the paths they need to reach critical breeding, feeding, and wintering grounds. If those corridors continue to be fragmented or interrupted by human activity, then the beauty and diversity of the natural world are sure to suffer in the future, as they are suffering in many places already.

Along the upper Mississippi, where spring and fall are again a celebration of sound and sight, as rafts of migrants return, a growing human population is threatening the habitat that the ducks, geese, cranes, and pelicans need to survive, both here and at other places along their annual routes. In the upper Midwest, more agricultural and urban development means fragmented grasslands, fewer wetlands, and more sediment and nutrients coursing into streams and rivers. Even utility wires can create a death trap for animals who travel highways in the sky. And birds once protected are now being hunted again.

In his classic celebration of the natural world of southwestern Wisconsin, *A Sand County Almanac,* Aldo Leopold wrote, "A thing is right when it tends to preserve the integrity, stability, and beauty of the biotic community. It is wrong when it tends otherwise." How to sustain a growing human population as part of the integrity and beauty of the biotic community, instead of a threat to it, has become one of the most pressing questions of the 21st century. Animals do not obey stop signs or move along new highways. They follow rain and food and their instinct to reproduce. Increasingly, the land along their migratory paths has been changed, interrupted, diminished.

Will the wildebeests continue to chase the rain across the open spaces of the Serengeti? Will white pelicans find their Gulf wintering grounds grimed with oil? Will great white sharks continue to prowl the open oceans? Will migration, the clock that ticks the tempo of the wilderness, beat on for future generations of animals—and the humans who need and admire them?

WITH A WHOOSH of wings and a splash, Canada geese pause, feed, rest, and then take off (opposite) to resume their migration toward northern breeding grounds. Young geese that hatched along the Mississippi River (above) are lifting off to begin the first migratory journey of their lives.

GREATER WHITE-FRONTED GEESE, snow geese, Canada geese, and dozens of other waterfowl species migrate in immense numbers through the Mississippi Flyway.

These geese found ample food at the Hultine Waterfowl Production Area in Nebraska, one of many wetland refuges that are managed for migrating water birds.

S eptember 1, 1914, was a day that symbolized the tragic excesses of human conduct against other animals. On that day in the Cincinnati zoo, the last passenger pigeon to grace the Earth died. Just two centuries before, these birds were "beyond number of imagination," a Virginia settler had written.

THE FUTURE OF MIGRATION

Their migrating flocks numbered literally millions of birds, so many that they eclipsed the sun as they flew by. The numbers themselves proved too tempting a target—passenger pigeons were just too easy to hunt, net, sell. In the mid-19th century, as big-city diners clamored for squab—young pigeons—hundreds of thousands were taken. At the same time, settlement took more and more of the forest habitats these migrating birds needed. The last confirmed sighting of a passenger pigeon in the wild was in 1900. Even an animal "beyond number of imagination" could disappear.

Today, legions of biologists, environmentalists, and concerned citizens are watching, monitoring, working—often against the tide of overexploitation, climate change, and habitat destruction—to save the migrating flocks and herds and pods of the natural world. In just a few decades the science and technology to do this work have become remarkably more sophisticated, and every year new inroads are being made and new insights are being discovered.

Following and understanding the behavior and biological engineering behind animals that don't have the limitations of humans is a challenge. How can scientists study

THE WHALE SHARK comes to the surface, perhaps to bask in the sun or because the microorganisms in its diet rise with upwellings from lower depths.

migrating birds, some of which travel many thousands of miles across the globe, and often at night? In the past, most of what we knew about their behavior came from direct observation, usually on their breeding or wintering grounds. To try to determine where their travels took them, individual birds had to be marked somehow, and the techniques to do that were fairly primitive, especially to the 21st-century eye.

As long ago as 1595 a metal band attached to a peregrine falcon of French King Henry IV was recovered in Malta. The process of banding hasn't changed radically since then, and it's still used by governments, scientists, and conservation groups to ascertain the movements of birds. But banding is labor intensive and provides limited data. The technology behind migration science has a lot more to choose from these days.

One technological leap forward occurred with the use of sonar and radar during World War II. The sonar used to detect enemy submarines gave science its first clue of the great daily migration of the deep scattering layer, whose thousands of zooplankton nightly move up through the ocean layers toward the surface to feed. Meanwhile, mid-century radar operators scanning the skies picked up unexplained interferences they called "angels." The angels, in fact, were flocks of migrating birds. Today's more sophisticated radar can do much more than simply detect angels—it can provide the altitude, speed, and wing beats of distinct birds in a passing flock.

Another tool that has proved invaluable to wildlife studies is our own ability to take to the air. The aerial surveys of East African wildlife by Bernhard and Michael Grzimek in the late 1950s broke new ground, and since then scientists and conservationists have refined aerial techniques, searching the oceans for elusive whales and walruses and making transects that cover terrain systematically. It was on just such a transect of Southern Sudan in 2008 that J. Michael Fay and Paul Elkan spotted the remarkable migration of more than a million gazelles and antelope—a migration whose numbers rival that of the Serengeti herds.

At about the time the Grzimeks were at work in Africa, other wildlife biologists were beginning to use radiolocation telemetry. At first the radio collars they attached to migrating animals had a narrow bandwidth, but in the next decade, very high frequency (VHF) collars revolutionized the study of animals on the move by greatly extending the range and detail of the data collected. But the revolution wasn't over. VHF was followed by satellite telemetry, which didn't require humans in the field to pick up the radio signals. Instead, the signals could be beamed to a satellite collection system called Argos and the data ultimately downloaded to a private computer via the Internet. More great leaps soon followed. By 1990 the Global Positioning System (GPS), developed by the United States Department of Defense for military purposes, began to be used for animal tracking. Combined with data on landscapes from GIS—the Geographic Information System—biologists sitting at their office computers could monitor where animals were and what kinds of landscapes they were passing through.

Still, collaring and tracking a caribou is generally an easier proposition than trying the same thing with a manatee, or a whale. Just getting to marine life to apply the collar is more difficult and requires a diver or snorkeler with skill and finesse. And the tracking devices themselves are likely to malfunction because of exposure to salt water or the wear and tear they're subjected to. Not infrequently, they become detached or the batteries need changing, again a delicate operation with a fish or marine mammal. One ingenious way around that is the use of PATs—pop-up archival tags—set to detach at a given time and float to the surface.

And then there is the newest innovation, by aquatic biologist Rory Wilson. Frustrated that he could really study penguins' behavior only when they were visible, he developed something he compares to a plane's black box recorder—but for a marine animal. He calls the matchbox-size device a "daily diary," and its minute instrumentation, packed in resin, records an animal's respiration rate, speed, depth, pitch, roll, and direction eight times a second.

The daily diary has revealed astonishing details of animal behavior. With it, Wilson can "see" how the penguins function in their natural element, the sea, and he's discovered that they calculate how much air they'll need for the most efficient use of energy as they dive. "Using the breath to submerge and ascend while catching prey is a brilliant way to conserve calories," says Wilson, a National Geographic grantee.

Tracking the underwater movements of elephant seals with the daily diary, Wilson realized that they were "behaving like birds. They are flying underwater. . . . Small birds flap their wings to gain momentum when flying up, and hitch a ride on the current going down." And that's what elephant seals were doing.

Small birds and other flying creatures pose their own special challenges to the biologists trying to follow them. Take the monarch butterfly, an animal weighing less than a hundredth of an ounce that makes a migration of as much as 2,000 miles. Until very recently, scientists were using the paper tag—the same tracking device that led entomologist Fred Urquhart to the monarchs' wintering grounds in

KING PENGUINS, the second largest penguin species (bested in size only by the emperor), spend much of the year traveling around the Southern Ocean before coming ashore in huge numbers to breed. The conservation organization BirdLife International estimates the world's king penguin population at two million birds, which breed widely on sub-Antarctic islands.

Mexico almost 40 years ago. But migration ecologist Martin Wikelski, working with biologist Chip Taylor, was determined to try electronic tracking technology on the fragile creature. He and his team had already designed a simple device—a tiny hearing-aid battery attached to an aluminum antenna—and had used it successfully on dragonflies and bees. With the help of a little Super Glue, Wikelski applied it to monarchs—and it worked. "Now we can compare the migrations of whales, birds, bats, and insects, and describe trends," Wikelski says.

Even the subtle clues of chemistry have been used in recent years to track the course of migrating animals. By matching the stable isotope deuterium in the animals' cells with vegetation at a particular location, scientists can get a better idea of the migratory path.

Aside from studying where animals migrate to and from, there is the question of how. Many migrating animals seem to have some sophisticated navigational equipment in common. Ecologist and evolutionary biologist James Gould defines navigation as the "neural processing of sensory inputs to determine direction and perhaps distance." And what's clear to him is that most migrating animals rely on several sensory inputs to do that. "Animals whose lives depend on accurate navigation are uniformly overengineered," Gould says.

Most appear to use sight cues and their memory of landscape markers to guide them on their migratory route. Beyond the obvious visual cues, many animals are able to see light that humans can't—ultraviolet, polarized, infrared—and they use these light cues, as well as a "sun compass," to guide them. Even on cloudy days or before the sun rises and after it sets, polarized light can keep the migrants headed in the right direction.

Stars, too, light the way for many animals, particularly birds, which apparently have a star chart in their heads. What is most remarkable about this is their ability to make adjustments in their flight patterns as they head in any direction, recalibrating their direction to correct for the Earth's declination, and the change in the night sky as the seasons and the animals' latitudinal positions change.

Odor, as well, may come into play with migrating animals, familiarizing them with local conditions in much the way that visual landmarks do. And for marine animals, wave action, temperatures, and currents can provide clues. Does that explain how loggerhead turtles find their way across 7,000 miles of ocean to return to their nesting beaches? Maybe to some extent, but recent studies seem to indicate that the turtles, like many other animals, are equipped with their own biological magnetic compass. Scientists have known for several decades that monarch butterflies, birds, and other migrating animals actually have microscopic bits of magnetite—a magnetic ore—in their bodies that sensitize them to the geomagnetic field. They "read" the Earth's geomagnetism and set their courses accordingly.

Even these navigational tools can't protect the Earth's migrants from the new and immediate threats they face—their overexploitation by hunters and fishers, the loss of their habitat, the human-made obstacles they find blocking their age-old routes, and a changing climate. "The phenomenon of migration is disappearing around the world," warns ecology and evolutionary biologist David Wilcove. And with migration goes the disappearance of species and the very web of biodiversity.

A number of organizations are working hard to protect migration corridors and reserves for the long-distance travelers. That will require thoughtful planning instead of rampant development or unbridled agriculture. It will require more public awareness and more involvement from and among governments, communities, and corporations.

But for the long-haul migrants—the white sharks and sea turtles, the arctic terns and pronghorn—the borders of nations and reserves hold no meaning. Nor, of course, do laws that might protect them and their environment in one country but not another. They move as they always have, with no thought to human rules and arrangements. For decades international efforts have been made to provide global protections—against landscape and climate degradation, against excessive hunting, fishing, and other abuses—and some of these efforts have been effective. Standards agreed upon by the International Whaling Commission have helped bring back cetacean populations that had been hunted to near extinction in previous decades. But international standards require complicated cooperation and coordination and the teeth of enforcement to make them effective, and that's often missing.

"Only within the moment of time represented by the present century has one species—man—acquired significant power to alter the nature of his world," Rachel Carson wrote in the mid-20th century. She was both right and wrong. That moment has extended into another century, and now more than ever, humankind has the power to alter the nature of the world—its atmosphere and ocean chemistry, its contours and forests and the creatures that moment by moment continue their great migrations.

THE PRONGHORN, an iconic symbol of North America's vast High Plains, is a close relative of the antelope. Its population was reduced severely during the past century by hunting, but restrictions have allowed it to increase to an estimate of more than 500,000. Nevertheless, human activity is encroaching steadily on its habitat, and wildlife conservation agencies are studying the species intensively.

PRODUCER'S NOTEBOOK

The one thing I won't miss is phone calls in the middle of the night.

I bolt upright. It's 3 a.m. and once again the phone is ringing. My mind goes through the usual list of reasons I get calls at 3 a.m. these days: the plane isn't coming, the helicopter can't fly, the gear is lost, the animals have moved on.

Deafening static alerts me that it's a satellite phone call. Somewhere in the murk, I pick up the voice of producer James Byrne, calling from the scorching desert of Mali, West Africa. He's elated. He's just jumped off a helicopter, having filmed one of our holy grails—a sunrise aerial shot of migrating elephants, passing through an enormous crack in a stunning cliff face called "La Porte des Éléphants" ("The Elephants' Door").

James relays the thrilling news. His joy is palpable even from 4,000 miles away. It's a thrilling moment for our series—one of the signature shots we only dreamed of getting but, nonetheless, worked ourselves to the bone to achieve.

I hang up the phone and lie down. What a great call. And I realize that, when this series is over, the one thing I may miss the most is phone calls in the middle of the night . . .

National Geographic's *Great Migrations* was a three-year odyssey for everyone involved. As a team and as individuals, it was a migration of sorts. An adventure that challenged us, exhausted us, and, in the end, left us richer for the journey. Our three-year production called upon the finest wildlife filmmakers in the world to share their skills and push themselves to new extremes. As Dereck and Beverly Joubert wrote to me while filming the zebra migration in Botswana's broiling Makgadikgadi salt pans, "It is a harsh, harsh place where the wind blows salt-laced sand at you daily. The grass looks wonderfully soft, but sit on it, and it rejects you with spikes and biting ants. Stepping out onto the salt pans is like a visit to Dante's inferno, and if you are less than robust, and want shade, you won't find it here except lying under the truck (with the biting ants). It is, in short, paradise!"

From the first day, there was something different about this project. Perhaps it was the unprecedented scale of the production, and the tremendous resources committed to its success. This was the most ambitious project in the history of National Geographic Television and the National Geographic Channel, and we felt the weight of that responsibility and opportunity. But the entire staff of NGT and NGC readily accepted this challenge. Together, we racked up some impressive totals:

· 400,000+ miles traveled
· 800+ shooting days
· temperatures ranging from 20 below zero to 120 above Fahrenheit
· more than 200 hours filming from helicopters
· more than 100 hours filming from shark cages, and
· more than 400 hours filming while hanging in trees.

But looking beyond the numbers, everyone involved in *Great Migrations* toiled so hard because we understood that our project was beginning at a critical time: with our planet increasingly burdened by our human footprint. We recognized that our task was to document the timeless journeys of wildlife—but some of these journeys may be running out of time. This galvanized the team as we set forth to film "the most moving stories on Earth."

From Sudan to Siberia, from Australia to the Azores, from Peru to Palau, *Great Migrations* filmmakers endured every environmental challenge and film production snafu you can imagine. Every one of our 50-plus location teams came home with epic stories of National Geographic filmmakers' struggles, and ultimately successes, around the globe. From Andy Casagrande's tireless work with Serengeti cheetahs to Adam Ravetch's tales of swimming with the Arctic's Pacific walruses; from Mark Lamble's endless search for one million little red flying foxes ("Where do one million migrating bats hide?") to Neil Rettig's brilliant faux muskrat lodge that let him film shoulder-to-wing with migrating birds on the Mississippi River, every storyteller embraced the Society's core mission, "inspiring people to care about the planet."

And ironically, after three years of filming, the longest single production in NGT history, we learned that time is short. In the Falkland Islands, filmmakers Katie Bauer and Mark Smith were stunned to find how few rockhopper and gentoo penguins remained in once burgeoning colonies; in Mexico's mountains, Stephanie Atlas will never forget flying an ultralight aircraft over illegally logged forest, once home to countless monarch butterflies; in Borneo, John Benam and Jesse Quinn gathered rare footage of spectacular gibbons, now racing one step ahead of encroaching palm plantations; and in Mali,

James Byrne and Bob Poole filmed the last great herd of migrating desert elephants, now gingerly picking their way through newly built farms and villages.

In the final analysis, all of us on the National Geographic Television and National Geographic Channel production teams were inspired and sobered by our journey. We gained a deeper sense of wonder and inspiration after filming just a few of the countless creatures that must run, swim, and fly for their lives. We also earned a deeper understanding of our planet's fragility—and that we humans are migrants, too. We are a restless species, harboring an impulse to move that is a key to our success and our dominion over the planet. But how will humans and wildlife travel together into the future?

Three years ago, the *Great Migrations* team had brash ambitions: We wanted to fundamentally change viewers' response to migrating creatures. From the earliest days, we visualized that people who watched our series might wake up the next day and gaze across a field, over the sea, or up to the sky with a new reaction. If they saw a flurry of migrating creatures, they wouldn't simply pause and say, "Wow, isn't that beautiful!" Instead, they'd stop and say, "Wow, I'm rooting for you . . ."

We hope the images in our films and in this book serve as a kind of touchstone—reminding us that life exists only when we move together and survive as one.

—David Hamlin
National Geographic Headquarters
Washington, D.C.

CALL TO ACTION

The National Geographic Society is one of the world's largest non-profit scientific and educational organizations. Founded in 1888 to increase and diffuse geographic knowledge, the Society's mission is to "inspire people to care about the planet."

National Geographic's programs support the Society's mission by funding critical research while developing public outreach and education programs that highlight environmental challenges faced around the world. Many of the stories communicated through the Society's global media come from our investment in forward-thinking people and pioneering projects. *Great Migrations* reflects much of this work.

The Society's Mission Programs support more than 300 explorers and scientists each year. Mission Programs support programs that:

- explore through cutting-edge field research and science
- incubate future generations of talent and technology, and
- educate youth and adults to embrace the world around them.

The largest of National Geographic's grant-giving programs, the Committee for Research and Exploration primarily supports field-based scientific research around the world, within the context of National Geographic's mission. The Committee for Research and Exploration has supported more than 9,200 projects and expeditions worldwide. The Committee's experts award more than 200 grants each year. Many of these grants are awarded to seasoned scientists, researchers, and explorers who are leaders in their fields, applying novel methodologies to groundbreaking field research. Major fields of study include anthropology, archaeology, astronomy, biology, geography, geology, oceanography, and paleontology.

Examples of the more than 200 animal migration research projects supported by the Committee for Research and Exploration include the pioneering work of Frederick Urquhart, a leader in the area of monarch migration, in the 1970s; Frank Craighead, who used radiotransmitters and satellites to study grizzly bears and eagles in their habitats; and Barbara Block, who expanded our knowledge of the bluefin tuna and other species moving through the world's oceans.

A grant program funded by the Committee for Research and Exploration, the Conservation Trust supports both conservation fieldwork activities around the world and public education campaigns that yield creative solutions to global issues, connect conservation to daily life, and empower individuals to take action. Most grantees are in the early stages of their careers and show potential to become leaders in their fields. National Geographic supports them when they need it most, when other organizations are reluctant to fund the innovative ideas of scientists who are not yet well known.

Martin Wikelski is an example of the individuals supported by the Conservation Trust and featured in *Great Migrations*. Wikelski is leading the development of Move Bank, an unprecedented animal migration database that will facilitate the long-term comparison of previous animal movement data with new findings to show how climate change, altered landscapes, and other factors drive animal movement.

The National Geographic Expeditions Council is a grant program dedicated to funding exploration of largely unrecorded or little-known areas of the Earth as well as regions undergoing significant environmental or cultural change. Since its inception in 1998, the Expeditions Council has funded projects that span the entire spectrum of exploration and adventure. Through the great stories these projects generate, National Geographic hopes to foster a deeper understanding of the world and its inhabitants.

Expeditions Council grantee Rory Wilson, featured in *Great Migrations,* has employed new technology to study a wide range of marine animals, operating under varied conditions, from polar regions to the tropics. Although animal migrations are ultimately necessary for species survival, the costs associated with them are poorly understood. Wilson's research aims to determine the costs on these animals as they operate unhindered in the wild and to look for particular tricks used by individual species to enhance their survival.

The Expeditions Council and other grant-making bodies at the Society are supported by the generous donations of individuals, corporations, and foundations.

To learn more about how to support these programs, please visit:

http://www.natgeotv.com/migrations
http://www.nationalgeographic.org/field

ABOUT THE FILM

Move as millions. Survive as one. National Geographic Channel's *Great Migrations* gives the word "move" a whole new meaning. This seven-part global programming event takes viewers around the world on the arduous journeys that millions of animals undertake to ensure the survival of their species. Photographed from land and air, in trees and cliff-blinds, on ice floes and underwater, *Great Migrations* tells the formidable, powerful stories of many of the planet's species and their movements, while revealing new scientific discoveries with breathtaking high-definition clarity.

Discover the miraculous migrations of red crabs on Christmas Island; flying foxes in Australia; army ants in Costa Rica; wildebeests, zebras, and Mali elephants in Africa; great white sharks in the world's oceans; microscopic plankton and jellyfish in Palau; and the white-eared kob, as seen by the first film crew on the ground in Sudan in 25 years.

The beauty of these stories is underscored by new knowledge of these species' fragile existence and their life-and-death quest for survival in an ever changing world. The all-National Geographic *Great Migrations* team spent two and a half years in the field, traveling 420,000 miles in 20 countries and all seven continents to bring this spectacular first-of-its-kind production to television in fall 2010.

PRESENTED BY

For more information about Great Migrations ancillary products, visit www.natgeotv.com/migrations

Great Migrations, narrated by Alec Baldwin and featuring original music by Anton Sanko, is produced by National Geographic Television for the National Geographic Channel.

Great Migrations Production Team
David Hamlin, *Series Producer*
Eleanor Grant, *Senior Writer*
James Byrne, *Producer*
Katie Bauer, John Benam, and Alicia Decina,
  Coordinating Producers
Stephanie Atlas, *Associate Producer*
Jesse Quinn, *Series Production Coordinator*
Emmanuel Mairesse, *Editor, Episodes 1 and 4*
Salvatore Vecchio, *Editor, Episode 2*
Christine Jameson Henry, *Editor, Episode 3*
Michelle Manassah, *Production Manager*
Chesapeake Sacks and Teresa Neva Tate, *Researchers*

National Geographic Television Staff
Michael Rosenfeld, *President*
Kathy Davidov, *Executive Vice President, Production*
Keenan Smart, *Executive Producer, Natural History*
Anne Tarrant, *Senior Producer, Natural History*

Susan Lach, *Post Production Supervisor*
Braden McIlvaine, *Director, Post Production Operations*
Scott Wyerman, *Senior Vice President,*
  Standards and Practices
Todd Hermann, *Director, Research*

National Geographic Channel Staff
Char Serwa, *Executive Producer*
Juliet Blake, *Senior Vice President of Production*
Steve Burns, *Executive Vice President of Content,*
  NGC U.S.
Sydney Suissa, *Executive Vice President of Content,*
  NGC International

Principal Cinematography
John Benam
Andy Brandy Casagrande IV
Martin Dohrn
Graeme Duane
Mark Emery
Justine Evans
Evergreen Films
Wade Fairley
Richard Fitzpatrick

Richard Foster
Johnny Friday
David Hannan
Clint Hempsall
Jonathan Jones
Dereck Joubert
Mark Lamble
Alastair MacEwen
John Mans
Richard Matthews
Andy Mitchell
Shane Moore
Bob Poole
Adam Ravetch
Neil Rettig
Joe Riis
Rick Rosenthal
Andrew Shillabeer
Mark Smith

Primary Scientific Consultants
Martin Wikelski
Rory Wilson
Iain Couzin

BIBLIOGRAPHY

Adamczewska, Agnieszka M., and Stephen Morris. "Ecology and Behavior of *Gecarcoidea natalis,* the Christmas Island Red Crab, during the Annual Breeding Migration." *Biological Bulletin* (June 2001), 305–320.

Alderfer, Jonathan, ed. *National Geographic Complete Birds of North America.* National Geographic Society, 2006.

Baughman, Mel, ed. *Reference Atlas to the Birds of North America.* National Geographic Society, 2003.

Bingham, Mike. "Rockhopper Penguin." International Penguin Conservation Work Group, 2010. Available online at www.penguins/cl.

Bonner, Nigel. *Seals and Sea Lions of the World.* Facts on File, 1999.

Byers, John A. *American Pronghorn: Social Adaptations and the Ghosts of Predators Past.* University of Chicago Press, 1997.

Carson, Rachel. *Silent Spring.* Houghton Mifflin, 1962.

Department of Sustainability and Environment, State of Victoria, 2001. "About Flying-foxes." Available online at www.dse.vic.gov.au.

De Roy, Tui, Mark Jones, and Julian Fitter. *Albatross: Their World, Their Ways.* Firefly Books, 2008.

Dingle, Hugh. *Migration: The Biology of Life on the Move.* Oxford University Press, 1996.

Fay, J. Michael, Paul Elkan, Malik Marjan, and Falk Grossman. "Wildlife Conservation Society Aerial Surveys of Wildlife, Livestock, and Human Activity in and around Existing and Proposed Protected Areas of Southern Sudan, Dry Season 2007." Wildlife Conservation Society in Cooperation with the Government of Southern Sudan.

Forsberg, Michael. *Great Plains: America's Lingering Wild.* University of Chicago Press, 2009.

Fryxell, J. M., and A.R.E. Sinclair. "Seasonal Migration by White-eared Kob in Relation to Resources." *African Journal of Ecology* (March 1988), 17–31.

Garbutt, Nick. *Wild Borneo: The Wildlife and Scenery of Sabah, Sarawak, Brunei and Kalimantan.* MIT Press, 2006.

Gerrard, Jon M., and Gary R. Bortolotti. *The Bald Eagle: Haunts and Habits of a Wilderness Monarch.* Smithsonian Institution Press, 1988.

Glick, Daniel. "End of the Road?" *Smithsonian Magazine* (January 2007). Available online at www.smithsonianmag.com/science-nature/pronghorn.html.

Gordon, Jonathan. *Sperm Whales.* Voyageur Press, 1998.

Gotwald, William H., Jr. *Army Ants: The Biology of Social Predation.* Cornell University Press, 1995.

Grossman, Falk, Paul Elkan, Paul Peter Awol, and Maria Carbo Penche. "Surveys of Wildlife, Livestock, and Human Activity in and around Existing and Proposed Protected Areas of Southern Sudan, Dry Season 2008." Wildlife Conservation Society in Cooperation with the Government of Southern Sudan.

Grzimek, Bernhard. *Serengeti Shall Not Die.* E.P. Dutton and Co., 1959.

Ham, Anthony. "The Lost Herd." *Virginia Quarterly Review* (Winter 2010), 4–26.

Heyman, William D., Rachel T. Graham, Björn Kjerfve, and Robert E. Johannes. "Whale sharks *Rhincodon typus* aggregate to feed on fish spawn in Belize." *Marine Ecology Progress Series* 215 (May 2001), 275–282.

Hoare, Ben. *Animal Migration: Remarkable Journeys in the Wild.* University of California Press, 2009.

Holldobler, Bert, and Edward O. Wilson. *The Ants.* Harvard University Press, 1990.

Hughes, Janice M. *The Migration of Birds: Seasons on the Wing*. Firefly Books, 2009.

Jay, Chadwick V., and Anthony S. Fischbach. "Pacific Walrus Response to Arctic Sea Ice Losses." United States Geological Survey Report, 2008. Available online at purl.access.gpo.gov/GPO/LPS96746.

Kgathi, Dorothy K., and Mary C. Kalikawe. "Seasonal Distribution of Zebra and Wildebeest in the Makgadikgadi Pans Game Reserve, Botswana." *African Journal of Ecology* (April 1993), 210–219.

Klimley, A. Peter, and David G. Ainley, eds. *Great White Sharks: The Biology of* Carcharodon carcharias. Academic Press, 1996.

Loope, Lloyd L., and Pau D. Krushelnycky. "Current and Potential Ant Impacts in the Pacific Region." *Proceedings of the Hawaiian Entomology Society* (December 2007), 69–73.

National Aeronautics and Space Administration. "NASA Ice Images Aid Study of Pacific Walrus Arctic Habitats." (December 2006). Available online at www.nasa.gov/centers/ames/research/2006/walrus.html.

Payne, Junaidi, and Cede Prudente. *Orangutans: Behavior, Ecology, and Conservation*. MIT Press, 2008.

Poole, Robert M. "Heartbreak on the Serengeti." *National Geographic* (February 2006). Available online at ngm.nationalgeographic.com/ngm/0602/feature1.

Riedman, Marianne. *The Pinnipeds: Seals, Sea Lions, and Walruses*. University of California Press, 1990.

Robinson, Carlos J., and Jaime Gómez-Gutiérrez. "Daily Vertical Migration of Dense Deep Scattering Layers Related to the Shelfbreak Area Along the Northwest Coast of Baja California, Mexico." *Journal of Plankton Research* (1998), 1679–1697.

Scott, Jonathan. *The Great Migration*. Elm Tree Books, 1988.

Shoshani, Jeheskel. *Elephants: Majestic Creatures of the Wild*. Facts on File, 2000.

Sinclair, A.R.E., and M. Norton-Griffiths, eds. *Serengeti: Dynamics of an Ecosystem*. University of Chicago Press, 1979.

Strange, Ian. "The Falklands' Johnny Rook." *Natural History* (1986), 54–61.

Tickel, W.L.N. *Albatrosses*. Yale University Press, 2000.

Urquhart, Fred A. *The Monarch Butterfly: International Traveler*. Nelson-Hall, 1987.

Vardon, Michael, et al. "Seasonal Habitat Use by Flying-foxes, *Pteropus alecto* and *P. scapulatus* (Megachiroptera), in Monsoonal Australia." *Journal of Zoology* (2001), 523–535.

Ward, Carlton, Jr. "Restless Spirits." *Africa Geographic* (July 2007), 34–41.

Watson, Rupert. *Salmon, Trout, and Charr of the World: A Fisherman's Natural History*. Swan Hill Press, 1999.

Whitehead, Hal. *Sperm Whales: Social Evolution in the Ocean*. University of Chicago Press, 2003.

Wilcove, David S. *No Way Home: The Decline of the World's Great Animal Migrations*. Island Press, 2008.

Wildlife Conservation Society. "Massive Migration Revealed." Available online at www.wcs.org/new-and-noteworthy/massive-migration-revealed.aspx.

Williams, Tony D. *The Penguins*. Oxford University Press, 1995.

Wilson, Edward O. "Army Ants: Inside the Ranks." *National Geographic* (August 2006). Available online at ngm.nationalgeographic.com/2006/08/army-ants/moffett-text/1.

Yates, Steve. *The Nature of Borneo*. Facts on File, 1992.

Zimmer, Carl. "From Ants to People, an Instinct to Swarm." *New York Times*, November 13, 2007. Available online at www.nytimes.com/2007/11/13/science/13traff.html.

ABOUT THE AUTHORS

K. M. Kostyal, a former senior editor for *National Geographic* magazine and the National Geographic book division, has written and edited books and articles on a wide range of subjects. In the past year, she has edited major books on changing cultural and climatic conditions in the circumpolar Arctic and on the natural and human history of the Great Plains. She also authored *Abraham Lincoln's Extraordinary Era,* a joint publication of National Geographic's book division and the Abraham Lincoln Presidential Library; *1776: A New Look at Revolutionary Williamsburg*; and *Lost Boy, Lost Girl: Escaping Civil War in Sudan* for National Geographic children's books.

David Hamlin is an Emmy award–winning filmmaker. He is senior producer, special projects, for National Geographic Television and the series producer of *Great Migrations*.

GALLERY CAPTIONS

Introduction

page 1: Left to right: Snow geese, sockeye salmon, and red crabs.

2-3: Going to sea on the Antarctic Peninsula, gentoo penguins line up and quickly dive in together.

4: A gentoo penguin checks the water before taking the plunge.

8-9: Plains zebras thrash about in a tight herd in Botswana.

10-11: Spawning salmon dominate traffic in the Ozernaya River on the Kamchatka Peninsula, Russia.

12-13: Elephants cross Serengeti National Park in Tanzania.

14-15: Golden jellyfish float upward toward the life-giving sun on a lake in Palau.

Chapter 1

26-27: Wildebeests migrate across the plains in Maasai Mara National Reserve, Kenya.

28-29: Female red crabs on Christmas Island, Australia, migrate from the rain forest to the seashore to mate and spawn.

30-31: Wintering monarch butterflies fill the sky during a warm day in Michoacán, Mexico.

32-33: A sperm whale pod with a large calf migrates offshore of the Azores Islands in the eastern Atlantic.

Chapter 2

80-81: King penguins and elephant seals cover the beach at Saint Andrews Bay on South Georgia Island.

82-83: Army ants carry their larvae and food during nightly migration on Barro Colorado Island, Panama.

84-85: White-eared kob race across the plains of Southern Sudan.

86-87: Wild Atlantic salmon make their way upstream on the Gaspé Peninsula in Quebec, Canada.

88-89: Australia's little red flying foxes pour from their roosts every night to forage in the forest.

Chapter 3

150-151: Plains zebras rush across the plains in Maasai Mara National Reserve, Kenya.

152-153: A proboscis monkey, her infant holding tightly, makes a flying leap in the Bornean forest.

154-155: Off the coast of western Australia, small fish cluster around a whale shark, using it as shelter from predators.

156-157: A pronghorn herd streaks through the plains near Pinedale, Wyoming.

158-159: A walrus herd migrates through Arctic waters twice yearly, tracking seasonal food.

Chapter 4

216-217: A white shark swims behind a school of mullet near the Neptune Islands, South Australia.

218-219: Desert elephants of Mali migrate through Africa's arid Sahel region.

220-221: Golden jellyfish swim to the surface of a Palauan lake each day, seeking life-giving sunlight.

222-223: Thousands of sandhill cranes pause to rest and feed along the Platte River in Nebraska.

280-281: Tundra swans fly through a cloudy Minnesota sky during their long migration from the high Arctic to wintering grounds on the southeastern coast of the United States.

PHOTO CREDITS

Unless otherwise noted below, interior images are courtesy of National Geographic Television (NGT), which provided high-definition video from the documentary film for use in this companion volume.

2-3, Paul Nicklen; 4, Paul Nicklen; 8-9, Frans Lanting; 10-11, Randy Olson; 12-13, Anup & Manoj Shah; 17, Frans Lanting; 18, Joel Sartore; 21, Paul Nicklen; 23, Mitsuaki Iwago/Minden Pictures; 24 (LE), Beverly Joubert; 26-27, Anup & Manoj Shah; 28-29, John Hicks; 30-31, Ingo Arndt/Foto Natura/Minden Pictures; 32-33, Hiroya Minakuchi/Minden Pictures; 35, Anup & Manoj Shah; 40, Anup & Manoj Shah; 41, Anup & Manoj Shah; 43, Mitsuaki Iwago/Minden Pictures; 44-45, Mitsuaki Iwago/Minden Pictures; 49, David Hamlin; 53, Peter Arnold, Inc./Alamy; 56-57, John Hicks; 59, Jamie Dertz, National Geographic My Shot; 63, Ingo Arndt/Foto Natura/Minden Pictures; 64-65, Stephanie Atlas; 66, Jim Brandenburg/Minden Pictures; 71, Flip Nicklin/Minden Pictures; 75, Flip Nicklin; 76-77, Flip Nicklin; 77 (RT), Flip Nicklin; 79 (UP RT), James Byrne; 79 (UP LE), Martin Withers/Flpa/Minden Pictures; 80-81, Paul Nicklen; 82-83, Christian Ziegler/Minden Pictures; 84-85, George Steinmetz; 86, Paul Nicklen; 87, Paul Nicklen; 88-89, Raoul Slater/Lochman Transparencies; 94-95, Frans Lanting; 95 (RT), Frans Lanting; 96-97, Frans Lanting; 98, Paul Nicklen; 99 (LE), Paul Nicklen; 99 (RT), John Eastcott & Yva Momatiuk; 100-101, Paul Nicklen; 103, Frans Lanting; 104, Paul Nicklen; 109, Christian Ziegler/Minden Pictures; 112, Christian Ziegler/Minden Pictures; 114 (UP), Mark Moffett/Minden Pictures; 114 (LO), Mark Moffett/Minden Pictures; 115, Christian Ziegler/Minden Pictures; 116, Mark Moffett/Minden Pictures; 117, Mark Moffett/Minden Pictures; 118-119, Mark Moffett/Minden Pictures; 121, Ingo Arndt/Foto Natura/Minden Pictures; 124, all photos by James Byrne; 125, Chris Johns; 126-127, James A. Sugar; 129, Mark Conlin/Larry Ulrich Stock; 133, Michio Hoshino/Minden Pictures; 134 (LE), Paul Nicklen; 134-135, Paul Nicklen; 136, Michael S. Quinton; 137, Melissa Farlow; 139, KEO Films; 142, Piotr Naskrecki/Minden Pictures; 143 (LE), Tim Laman; 143 (RT), Tim Laman; 144-145, Roy Toft; 146-147, Lochman Transparencies; 150-151, Anup & Manoj Shah; 152-153, Tim Laman; 154-155, Brian Skerry; 156-157, Joel Sartore; 158-159, Paul Nicklen; 161, Robert B. Haas; 164-165, Richard Du Toit/Minden Pictures; 165 (RT), Mitsuaki Iwago/Minden Pictures; 166-167, Anup & Manoj Shah; 168 (LE), Robert B. Haas; 168-169, Marc Moritsch; 170-171, Beverly Joubert; 173, Colin Parker, National Geographic My Shot; 176 (LE), Kai Benson; 176-177, Brian Skerry; 178, Tim Laman; 179, Tim Laman; 181, Frans Lanting; 184-185, Tim Laman; 185 (RT), Cede Prudente/NGT; 186 (LE), Tim Laman; 186 (RT), Cede Prudente/NGT; 187 (LE), Tim Laman; 188, Tim Laman; 189, Tim Laman; 190 (LE), Mattias Klum; 190-191, Tim Laman; 193, Joe Riis; 196 (LE), Joe Riis; 196-197, Joel Sartore; 198, Joe Riis; 200 (LE), Joe Riis; 200-201, Joe Riis; 202-203, Michael Durham/Minden Pictures; 205, Joel Sartore; 209, Paul Nicklen; 210-211, Paul Nicklen; 211

(RT), Paul Nicklen; 212, Paul Nicklen; 213, Paul Nicklen; 216-217, Mike Parry/Minden Pictures; 218-219, Carlton Ward Jr.; 222-223, Joel Sartore; 225, Mauricio Handler; 228-229, Rich Reid; 230, Tim Fitzharris/Minden Pictures; 232-233, Michael Durham/Minden Pictures; 234, Brian Skerry; 235, Mauricio Handler; 237, Carlton Ward Jr.; 240, Carlton Ward Jr.; 242-243, Carlton Ward Jr.; 244-245, Carlton Ward Jr.; 246-247, Steve McCurry; 247 (RT), Carlton Ward Jr.; 254-255, Tim Laman; 255 (RT), Tim Laman; 261, Michael Forsberg; 264-265, Jim Brandenburg/Minden Pictures; 267, Pal Hermansen/Getty Images; 268-269, Tim Fitzharris/Minden Pictures; 270, Yva Momatiuk & John Eastcott/Minden Pictures; 271 (LE), Thomas Mangelsen/Minden Pictures; 273, Macduff Everton; 274 (LE), Klaus Nigge; 274-275, Annie Griffiths Belt; 276, Thomas Kitchin & Victoria Hurst/Getty Images; 280-281, Jim Brandenburg/Minden Pictures; 283, Paul Nicklen; 286, Joel Sartore; 291, Jim Brandenburg/Minden Pictures.

The montages on the pages noted below were created by blending several images together at their edges, seamlessly flowing from one photograph into the next, to produce a single illustration that conveys both the feeling of motion and the scale of the migration. The photographs themselves have not been manipulated in any way.

36-37 (L to R), Michael Poliza, Mitsuaki Iwago/Minden Pictures, Anup & Manoj Shah, Anup & Manoj Shah; 46-47 (L to R), Suzi Eszterhas/Minden Pictures, Suzi Eszterhas/Minden Pictures,Suzi Eszterhas/Minden Pictures, Anup and Manoj Shah, Anup and Manoj Shah, Chris Johns; 50-51 (L to R), John Hicks, John Hicks, National Geographic Television (NGT), Roger Garwood, David Hamlin; 54-55 (L to R), Frederique Olivier, John Hicks, NGT, NGT, NGT, NGT; 60-61, all photos by NGT; 68-69 (L to R), James L. Amos, Stephanie Atlas, Thomas Marent/Minden Pictures; 72-73 (L to R), Patricio Robles Gil/Minden Pictures, Flip Nicklin/Minden Pictures, Flip Nicklin; 92-93, all photos by Frans Lanting; 106-107 (L to R), Frans Lanting, Paul Nicklen; 110-111, all photos by Christian Ziegler/Minden Pictures; 122-123, all photos by Paul Elkan and Mike Fay; 130-131 (L to R), NGT, Larry Ulrich, Paul Nicklen; 140-141 (L to R), Konrad Wothe/Minden Pictures, Lochman Transparencies, KEO Films; 162-163 (L to R), Michael and Patricia Fogden/Minden Pictures, Anup and Manoj Shah; 174-175, all photos by NGT; 182-183 (L to R), Tim Laman, Mattias Klum, Tim Laman; 194-195 (L to R), Michael Durham/Minden Pictures, Patricio Robles Gil/Minden Pictures, Joe Riis; 206-207, all photos by NGT; 226-227 (L to R), Mauricio Handler, Brandon Cole/Visuals Unlimited/Getty Images, David Doubilet; 238-239 (L to R), Michael Fay, Carlton Ward Jr., Carlton Ward Jr.; 248-249, all photos by Carlton Ward Jr.; 252-253, all photos by NGT; 262-263, all photos by Michael Forsberg; 278-279 (L to R), Michael Forsberg, Sumio Harada/Minden Pictures, Michael Forsberg.

6'11
Bot

591.568
K05

GREAT MIGRATIONS

K. M. Kostyal

Published by the National Geographic Society

John M. Fahey, Jr., *President and Chief Executive Officer*
Gilbert M. Grosvenor, *Chairman of the Board*
Tim T. Kelly, President, *Global Media Group*
John Q. Griffin, *Executive Vice President; President, Publishing*
Nina D. Hoffman, *Executive Vice President;*
    President, Book Publishing Group

Prepared by the Book Division

Barbara Brownell Grogan, *Vice President and Editor in Chief*
Marianne R. Koszorus, *Director of Design*
Lisa Thomas, *Senior Editor*
Carl Mehler, *Director of Maps*
R. Gary Colbert, *Production Director*
Jennifer A. Thornton, *Managing Editor*
Meredith C. Wilcox, *Administrative Director, Illustrations*

Staff for This Book

Garrett Brown, *Editor*
Jane Menyawi, *Illustrations Editor*
Cameron Zotter, *Designer*
Sam Serebin, *Photo Montage Consultant*
Robert Waymouth, *Illustrations Specialist*
Paul Hess, *Picture Captions Writer and Copy Editor*
Scott Pospiech and Teresa Tate, *Researchers*
Steven D. Gardner and Gregory Ugiansky, *Map Research*
    and Production
Mike Horenstein, *Production Manager*

Manufacturing and Quality Management

Christopher A. Liedel, Chief Financial Officer
Phillip L. Schlosser, Vice President
Chris Brown, Technical Director
Nicole Elliott, Manager
Rachel Faulise, Manager

The Book Division extends a special thanks to David Hamlin,
Susan Lach, and Jesse Quinn at National Geographic Television
for their many contributions to this project.

The National Geographic Society is one of the world's largest nonprofit scientific and educational organizations. Founded in 1888 to "increase and diffuse geographic knowledge," the Society works to inspire people to care about the planet. It reaches more than 325 million people worldwide each month through its official journal, *National Geographic,* and other magazines; National Geographic Channel; television documentaries; music; radio; films; books; DVDs; maps; exhibitions; school publishing programs; interactive media; and merchandise. National Geographic has funded more than 9,000 scientific research, conservation and exploration projects and supports an education program combating geographic illiteracy. For more information, visit nationalgeographic.com.

For more information, please call 1-800-NGS LINE
(647-5463) or write to the following address:

National Geographic Society
1145 17th Street N.W.
Washington, D.C. 20036-4688 U.S.A.

Visit us online at www.nationalgeographic.com

For information about special discounts for bulk purchases, please
contact National Geographic Books Special Sales: ngspecsales@ngs.org

For rights or permissions inquiries, please contact National Geographic Books
Subsidiary Rights: ngbookrights@ngs.org

Copyright © 2010 National Geographic Society.
All rights reserved. Reproduction of the whole or any part of the contents
without written permission from the publisher is prohibited.

Library of Congress Cataloging-in-Publication Data

Kostyal, K. M., 1951-
Great migrations / by Karen M. Kostyal.
   p. cm.
Includes bibliographical references.
ISBN 978-1-4262-0644-3 (hardcover)
1. Animal migration. 2. Animal migration--Pictorial works. I. Title.
QL754.K67 2010
591.56'8--dc22

       2010020998

Printed in Italy

10/MV/1